AF368947

Applications of Biosorption in Wastewater Treatment

Applications of Biosorption in Wastewater Treatment

Editor

José Enrique Torres Vaamonde

MDPI • Basel • Beijing • Wuhan • Barcelona • Belgrade • Manchester • Tokyo • Cluj • Tianjin

Editor
José Enrique Torres Vaamonde
Universidade da Coruña
Spain

Editorial Office
MDPI
St. Alban-Anlage 66
4052 Basel, Switzerland

This is a reprint of articles from the Special Issue published online in the open access journal *Processes* (ISSN 2227-9717) (available at: https://www.mdpi.com/journal/processes/special_issues/biosorption_wastement_treatment).

For citation purposes, cite each article independently as indicated on the article page online and as indicated below:

LastName, A.A.; LastName, B.B.; LastName, C.C. Article Title. *Journal Name* **Year**, *Volume Number*, Page Range.

ISBN 978-3-0365-3533-3 (Hbk)
ISBN 978-3-0365-3534-0 (PDF)

Contents

About the Editor

José Enrique Torres Vaamonde is currently working at the Microbiology Laboratory of the Faculty of Sciences of the University of Coruña (Spain) as an Associate Professor. He received his PhD in Biology from University of Coruña in 1997. His main scientific focus is on the study and application of systems based on microalgae. He has more than 25 years of experience in the application of these microorganisms for various scientific purposes, including in studies of the toxic effects of pollutants, studies of response mechanisms and tolerance to pollutants, pollutant removal and physiological studies. He has published more than 45 papers in international peer-reviewed journals and has reviewed papers in various international scientific journals. He has supervised several PhD theses and has participated in several projects.

MDPI

Editorial

Special Issue "Applications of Biosorption in Wastewater Treatment"

José Enrique Torres Vaamonde

Department of Biology, Faculty of Sciences, Universidade da Coruña, 15008 A Coruna, Spain; enrique.torres@udc.es

Citation: Vaamonde, J.E.T. Special Issue "Applications of Biosorption in Wastewater Treatment". *Processes* **2022**, *10*, 22. https://doi.org/10.3390/pr10010022

Received: 22 November 2021
Accepted: 24 November 2021
Published: 23 December 2021

Publisher's Note: MDPI stays neutral with regard to jurisdictional claims in published maps and institutional affiliations.

The interest in the use of biosorption for the elimination of pollutants is because this technique is an efficient and environmentally friendly process, constituting an alternative to the so-called conventional treatment processes. Low-cost or disposable materials of biological origin (biomass) are used, which allows these materials to be useful. For these reasons, the current contributions in the field of biosorption continue to be numerous and very attractive. An example is this Special Issue on the "Applications of Biosorption in Wastewater Treatment" in the journal *Processes*, which is available online at: https://www.mdpi.com/journal/processes/special_issues/biosorption_wastement_treatment (accessed on 25 November 2021).

In this Special Issue, several articles are published that demonstrate the high efficiency of this process, its versatility to develop different applications, and show the recent advances. Thus, the review by Torres [1] shows the recent contributions made in the field of pollutant removal using biosorption. The general aspects of, and the usefulness of this process are included in this review. The process can be applied to most pollutants, regardless of their nature, which demonstrates its operability. This article shows numerous examples of this process as a solution to the elimination of pollutants, which demonstrates the scientific interest in developing applications using biosorption. This variety of examples is due to the fact that the sorbent material of biological origin has a varied nature, which allows the evaluation of a large number of materials, in order to find the most suitable for a certain application. The use of living biomass is a possible option in this field when biological material is applied, preserving its activity. It is an excellent alternative that can be used in different situations, and offers a great efficiency, as shown in this article.

One of the great advantages of the biosorption process is that the sorbent material can be very varied; thus, it is possible to use waste materials that lack an apparent utility. As an example of this, we have the article by Mahato et al. [2] in this Special Issue. This review shows the application of biomass from citrus waste for the elimination of a great diversity of pollutants. This waste, produced in large quantities around the world, can be transformed into an excellent biosorbent. The article analyzes its synthesis and sorption capacity in detail, and also provides information on its biotransformation for the production of biofuel and other valuable compounds.

In biosorption studies, it is common to consider the effect of different parameters on the performance of the process. Thus, there are several parameters that influence the performance of the biosorption process, such as pH, temperature, type of biomass, and nature of the sorbent, to name some of the most studied. However, there are other lesser-known factors that also influence the process, such as ionic strength. An interesting study by Aranda-García et al. [3], published in this Special Issue, shows the inversely proportional effect of ionic strength on nickel biosorption capacity. In addition, the effect of this capacity of different background electrolytes, and how these can alter the performance of the process, is also studied. This is interesting because pollutants from real solutions are not dissolved in distilled water, which means that these solutions may contain other components that influence the performance of the pollutant biosorption process. These authors used acorn shell from the oak *Quercus crassipes* as biomass, and demonstrated that this biomass was

1

suitable for removing nickel ions from solutions containing impurities at different concentrations. Few studies have included the effect of ionic strength; however, in this Special Issue, this parameter was also considered in the article by Villen-Guzman et al. [4]. This article studies the ability of biomass from *Spirulina* to remove lead ions. This is not only a typical biosorption article: it also includes a study with this biomass immobilized in alginate. The immobilization of biomass is another step in the improvement of biosorption processes, since it provides mechanical strength, rigidity, ideal size, and porous characteristics. The promising results obtained in this study for immobilized *Spirulina* on alginate beads, could be the first step towards the valorization of this biomass for the treatment of industrial wastewater polluted by toxic metals, such as lead.

Immobilization is not the only tool to improve the biosorption process. The combination of biomass with modern materials to form composites is another very useful tool for this purpose. This combination allows a greater stability of the biomass, which increases the performance of the process. An example is found in the article by Wan et al. [5] in this Special Issue, in which a composite consisting of Mxene (transition metal carbide, nitride, or boride) and chitosan that was used to remove chromium ions, is successfully obtained. This novel Mxene–chitosan composite can be considered as an alternative for the adsorption of heavy metals from wastewater.

The application of biomass for the removal of pollutants from wastewater goes beyond biosorption, since biomass can also act as a substrate to promote the biodegradation of organics in wastewater. Banch et al. [6] used palm oil mill effluent to treat stabilized leachate from old landfill. The biological treatment of this mixture with an aeration process was able to remove the organics effectively.

To conclude, this Special Issue shows that biosorption constitutes a very useful tool for the removal of pollutants from wastewater. The variety of articles published, where different modalities of biosorption are exposed, demonstrates the enormous versatility of this technique, and allows us to continue advancing in a promising future in the field of biosorption. The state-of-the-art applications of biosorption presented in this Special Issue may serve as valuable references for future research in this ever-evolving field.

Thanks goes to all the authors who contributed to the development of this Special Issue and the editorial staff of *Processes* for their efforts.

Funding: There are no funding supports.

Conflicts of Interest: The author declares no conflict of interest.

References

1. Torres, E. Biosorption: A Review of the Latest Advances. *Processes* **2020**, *8*, 1584. [CrossRef]
2. Mahato, N.; Agarwal, P.; Mohapatra, D.; Sinha, M.; Dhyani, A.; Pathak, B.; Tripathi, M.K.; Angaiah, S. Biotransformation of Citrus Waste-II: Bio-Sorbent Materials for Removal of Dyes, Heavy Metals and Toxic Chemicals from Polluted Water. *Processes* **2021**, *9*, 1544. [CrossRef]
3. Aranda-García, E.; Chávez-Camarillo, G.M.; Cristiani-Urbina, E. Effect of Ionic Strength and Coexisting Ions on the Biosorption of Divalent Nickel by the Acorn Shell of the Oak *Quercus crassipes* Humb. & Bonpl. *Processes* **2020**, *8*, 1229. [CrossRef]
4. Villen-Guzman, M.; Jiménez, C.; Rodriguez-Maroto, J.M. Batch and Fixed-Bed Biosorption of Pb (II) Using Free and Alginate-Immobilized Spirulina. *Processes* **2021**, *9*, 466. [CrossRef]
5. Wan, H.; Nan, L.; Geng, H.; Zhang, W.; Shi, H. Green Synthesis of A Novel MXene–CS Composite Applied in Treatment of Cr(VI) Contaminated Aqueous Solution. *Processes* **2021**, *9*, 524. [CrossRef]
6. Banch, T.J.H.; Hanafiah, M.M.; Amr, S.S.A.; Alkarkhi, A.F.M.; Hasan, M. Treatment of Landfill Leachate Using Palm Oil Mill Effluent. *Processes* **2020**, *8*, 601. [CrossRef]

Article

Treatment of Landfill Leachate Using Palm Oil Mill Effluent

Tawfiq J. H. Banch [1], Marlia M. Hanafiah [1,2,*], Salem S. Abu Amr [3], Abbas F. M. Alkarkhi [3] and Mohammed Hasan [1]

[1] Department of Earth Sciences and Environment, Faculty of Science and Technology, Universiti Kebangsaan Malaysia, Bangi 43600, Selangor, Malaysia; banch4@gmail.com (T.J.H.B.); aldulaimi89_fst@yahoo.com (M.H.)
[2] Centre for Tropical Climate Change System, Institute of Climate Change, Universiti Kebangsaan Malaysia, Bangi 43600, Selangor, Malaysia
[3] Universiti Kuala Lumpur Business School (Unikl bis), Kuala Lumpur 50250, Malaysia; sabuamr@hotmail.com (S.S.A.A.); alkarkhii@gmail.com (A.F.M.A.)
* Correspondence: mhmarlia@ukm.edu.my

Received: 18 April 2020; Accepted: 14 May 2020; Published: 18 May 2020

Abstract: Sanitary landfilling is the most common method of removing urban solid waste in developing countries. Landfills contain high levels of organic materials, ammonia, and heavy metals, thereby producing leachate which causes a possible future pollution of ground and surface water. Recently, agricultural waste was considered a co-substratum to promote the biodegradation of organics in industrial wastewater. The use of low-cost and natural materials for wastewater treatment is now being considered by many researchers. In this study, palm oil mill effluent (POME) was used for treating stabilized leachate from old landfill. A set of preliminary experiments using different POME/leachate ratios and aeration times was performed to identify the setting of experimental design and optimize the effect of employing POME on four responses: chemical oxygen demand (COD), total suspended solids (TSS), color, and ammoniacal nitrogen (NH_3-N). The treatment efficiency was evaluated based on the removal of four selected (responses) parameters. The optimum removal efficiency for COD, TSS, color, and NH_3-N was 87.15%, 65.54%, 52.78%, and 91.75%, respectively, using a POME/leachate mixing ratio of 188.32 mL/811.68 mL and 21 days of aeration time. The results demonstrate that POME-based agricultural waste can be effectively employed for organic removal from leachate.

Keywords: landfill; leachate treatment; POME; removal efficiency; mixing ratio

---

1. Introduction

The landfilling of solid waste is still a significant problem in the solid waste management systems of all countries worldwide [1,2]. However, landfill leachate is a complex liquid generated from rainwater penetration through landfills that often includes high-strength contaminant resistance, such as humic acids, ammonia nitrogen, heavy metals, xenobiotics, and inorganic salts, which are important to avoid due to their adverse effects on the environment [3,4]. The factors that affect the composition of landfill leachate include the composition of waste, the level of compaction, the absorptive capacity of solid waste and age of solid waste, weather variations, precipitation, landfilling temperature, size of landfilling, hydrogeological conditions, factors of the landfill operation, pH, and chemical and biological activities in the process of landfilling [5,6]. In general, young leachate produced from new landfills (<5 years old) has a large biodegradation of comparatively low-molecular-weight materials, such as volatile organic acids, chemical oxygen demand (COD), total organic carbon (TOC), biological oxygen demand (BOD_5), and biodegradability (BOD_5/COD) [7]. On the other hand, old leachate (>10 years old) has high organic content of relatively high-molecular-weight materials, for instance, humic and fulvic substances that are refractory and not rapidly degradable. Generally, old leachate has a lower concentration of COD, TOC, BOD_5, and BOD_5/COD [6,7]. Leachate is stabilized and has low biodegradability since most landfills are old [8–11]. As a landfill gets older, a change from a relatively shorter initial aerobic to a longer anaerobic decomposition period takes place. Due to the biological breakdown of organic compounds and precipitation of soluble components, such as heavy metals, the strength of leachate generally lowers over time. Because of its biodegradable nature, organic compounds decrease faster than inorganic compounds with the increasing age of leachate production.

Several treatment techniques, such as physico-chemical processes, are used to treat leachate (coagulation precipitation, activated carbon adsorption [12,13], membrane filtration, activated carbon adsorption [14,15], and/or other separation techniques) [16], in addition to biological treatment methods, such as aerobic and anaerobic processes. In the past two decades, many studies reported that biological processes in the treatment of young leachate are effective but are comparatively insufficient in the treatment of old leachate because of the existence of bio-refractory substances [6,7]. Currently used procedures often involve mixed methods intended as modular or multi-stage units that are capable of treating pollutants that change in concentration over time. The appropriate treatment of leachate is applied to enhance and develop an appropriate technique of treatment that meets the relevant quality standards and regulations [5].

In particular, the palm oil sector in Malaysia adopts four types of treatment technologies: waste stabilization ponds, activated sludge systems, closed anaerobic digesters, and land application systems [17]. However, many questions remain unanswered about the cost of treatment, the production of sludge, and chemical residues in treated wastewater. Usage of low-cost and natural materials of wastewater treatment are currently gaining increased attention from a series of studies. Palm oil mill effluent (POME) is classified as a pollutant which is known for its ability to enhance the biodegradability of pollutants due to the relatively high content of organic matter [18,19]. An aeration process supplies the oxygen to the wastewater and acts as an oxidation of organics. Moreover, oxygen can also enhance the biodegradation of organics by bacteria which use oxygen to break down the organic matter into the form of CO_2 and H_2O [20]. Stripping of ammonia is a simple desorption method utilized to reduce the ammonia concentration of leachate. Ammonia reacts with water to form ammonium hydroxide [21], as shown in Equation (1).

$$NH_4^+ + OH^{-*} \leftrightarrow H_2O + NH_3^*. \tag{1}$$

The alkaline wastewater flows downward when the air enters through the bottle. As the air continues to flow, the wastewater moves in the opposite direction and ascends to the top of the aeration bottle. NH_3 is stripped from dropping water droplets into the air flow, and then released into the atmosphere [22]. In this research, the effectiveness of aeration processes used in the treatment of a leachate/POME mixture was evaluated. The optimum experimental conditions for POME dosages were designed, and the ability to remove the different medicines was studied. In addition, the efficacy of POME in the leachate removal of heavy metals was also investigated. The removal of heavy metals by POME during aeration may be attributed to the high level of suspended solids in POME, which may act as a natural coagulant [23].

2. Methodology

2.1. Collection of the Samples and Landfill Characteristics

Samples of leachate were sampled from the Ampar Tenang Closed Landfill Site (ATCL). ATCL is situated nearly 40 km southwest of Kuala Lumpur, Malaysia at a latitude of 02°48.9250 north (N) and a longitude of 101°4.9330 east (E) [23]. ATCL is mostly surrounded by oil palm plantations. Labu River flows along the landfill 300 m away. The ATCL area is characterized as tropical. The average temperature is 27.2 °C, and the average rainfall is 2287 mm/year [24]. The landfill site is located on the eastern sector of the confined alluvial aquifer of the Langat Basin, composed primarily of silt (50%–70%), clay (<25%), and sand (<25%) [25–27]. This site is more clayey close to the surface of the ground, but changes to sandy in lower layers [25,26]. ATCL operated as of 1994 for a total of 15 years. About 100 tons of waste are dumped into this site every day during the operation of the landfill. This results in the on-site disposal of a total of half a million tons of solid waste. The site was completely closed in 2010. Before it was closed, it was converted to sanitary classification (Level 1) from a disposal site (Level 0) [8,27,28]. In 2018, the samples were manually collected and placed in 1000-mL containers manufactured from polyethylene. The samples were collected at 4 °C and immediately transformed to prevent significant biological degradation and chemical reactions. The samples were transported in a preservation refrigerator from the site to the laboratory before being stored in the laboratory refrigerator until the next day where the experiment was prepared; no acids were used for preservation.

2.2. Leachate and POME Characterizations

The physico-chemical characterization of the leachate is presented in Table 1.

Table 1. Physico-chemical characteristics and heavy metals of leachate.

Parameter	Mean and Standard Deviation	(USEPA *; DOE **)
pH	7.88 ± 0.50	6–9 **
EC (μS/cm)	6565 ± 324	1400 *
TDS (mg/L)	4671 ± 174	1000 *
TSS (mg/L)	40.45 ± 8	50 **
COD (mg/L)	893.41 ± 202	400 **
BOD_5 (mg/L)	59.20 ± 10	20 **
NH_3-N (mg/L)	530.7 ± 22	5 **
DO (mg/L)	5.43 ± 2	5 *
Mg^{2+} (mg/L)	19.72 ± 4	0.2 *
Ca^{2+} (mg/L)	39.72 ± 34	0.1 *
Na^+ (mg/L)	638.80 ± 303	0.02 **
Fe^{2+} (mg/L)	0.78 ± 0.6	5 **
Zn^{2+} (μg/L)	280 ± 16	2 **
Cu^{2+} (μg/L)	41.87 ± 24	0.2 **
Cr^{2+} (μg/L)	45.11 ± 17	0.01 **
Cd^{2+} (μg/L)	0.62 ± 0.7	0.01 **
Pb^+ (μg/L)	4.18 ± 2.7	0.01 **
As^{3+} (μg/L)	17.3 ± 7	0.05 **
Co^{2+} (μg/L)	11.05 ± 8	0.05 *
Mn^{2+} (μg/L)	61.40 ± 49	0.02 **

* [29], ** [30]. Abbreviation. EC: electrical conductivity; TDS: total dissolved solids; TSS: total suspended solids; COD: chemical oxygen demand; BOD_5: biochemical oxygen demand; NH_3-N: ammoniacal nitrogen; DO: dissolved oxygen; USEPA: U.S. Environmental Protection Agency; DOE: Department of Environment.

The physico-chemical characterization of the POME is presented in Table 2.

Table 2. Physico-chemical characteristics and heavy metals of palm oil mill effluent (POME).

Parameter	Mean and Standard Deviation	(Standard of DOE *)
pH	4.40 ± 0.01	5.0–9.0
EC (μS/cm)	8553 ± 114	100
Salinity (ppt)	4.9 ± 0	NA
TDS (mg/L)	5713 ± 15	NA
TSS (mg/L)	3483 ± 76	200
Color (Pt/Co)	5517 ± 104	100
COD (mg/L)	$17,400 \pm 100$	NA
BOD_5 (mg/L)	1243 ± 51	NA
BOD_5/COD	0.07 ± 0	NA

Table 2. *Cont.*

Parameter	Mean and Standard Deviation	(Standard of DOE *)
NH_3-N (mg/L)	308 ± 58	NA
DO (mg/L)	6.74 ± 0.05	NA
Mg^{2+} (mg/L)	285 ± 5	NA
Ca^{2+} (mg/L)	47.34 ± 0.03	NA
Fe^{2+} (mg/L)	45.31 ± 0.9	50
Zn^{2+} (µg/L)	2603 ± 5.77	10
Cu^{2+} (µg/L)	2130.00 ± 26.46	10
Cr^{2+} (µg/L)	910.00 ± 10	NA
Cd^{2+} (µg/L)	90.10 ± 0.03	NA
Pb^{+} (µg/L)	54.38 ± 0.31	NA
As^{3+} (µg/L)	100.22 ± 0.44	NA
Co^{2+} (µg/L)	41.08 ± 0.17	NA
Mn^{2+} (µg/L)	80.15 ± 0.26	10

* DOE: Department of Environment, Malaysia; NA—not available [30].

2.3. Experimental Procedure

A bubble column bioreactor was used for the aeration process. It is characterized by its simple construction, higher efficiency in removal, and efficient control of the liquid residence time [31]. The aeration process was conducted at room temperature (25 °C). In the aeration process, the air pump is connected to four bottles of one liter each by a tube of 1 mm in diameter; these bottles contain different ratios of leachate/POME, and the process continues for 21 days.

The experiment was conducted in two steps; the first step was a preliminary experiment performed utilizing one factor at a time to determine the area of concern for each influential variable of the leachate/POME ratio and the aeration time to determine the optimal levels. The selected levels for the leachate/POME ratio and the aeration time were utilized to conduct the second step utilizing response surface methodology (RSM). RSM consists of a group of experimental methods devoted to estimating the relationship between a group of experimental variables (factors) and the (targeted) measured responses. To build a more practical model, the process variables under investigation need to be understood. Central composite face-centered (CCF), a type of central composite design (CCD), was used for two independent variables to estimate the effect value of POME dosages and aeration time on four response variables: COD, TSS, color, and NH_3-N.

2.4. Effect of the Leachate/POME Ratio

For the first step, POME was used to improve the biodegradation of leachate. In 1000 mL of leachate samples, different leachate/POME ratios (1:0, 0.9:0.1, 0.7:0.3, and 0.5:0.50) were used. The initial pH for the leachate sample (8.4) was left unadjusted. The liquid was aerated using an aeration bump (HAILEA) model V-20 with output 20 L/min and pressure >0.02 MPa for 24 days. The treatment efficiency was evaluated based on COD, TSS, color, and NH_3-N removal efficiency.

The efficiency for COD removal was estimated using Equation (2).

$$COD \text{ Removal } (\%) = [(C_r - C_k)/C_i] \times 100, \qquad (2)$$

where C_r is the initial COD concentration, and C_k is the final COD concentration.

2.5. Optimization of Treatment Efficiencies of Targeted Parameters

Using Design-Expert software (version 6.0.7), a central composite design (CCD) for the leachate/POME ratio was developed to examine whether COD, TSS, color, and NH_3-N affected the leachate/POME ratio and aeration time. Depending on the preliminary experiments stated in Section 2.3, the amounts and rates of each factor were chosen. Thirteen experiments were conducted to include all possible combinations of the leachate/POME ratio and aeration time.

Data from different CCD experiments were utilized to appropriate a polynomial model and a second-order model (Equation (3)).

$$Y = \beta_0 + \sum_{j=1}^{k} \beta_j X_j + \sum_{j=1}^{k} \beta_{jj} x^2_j + \sum_{i} \sum_{<j=2}^{k} \beta_{ij} X_i X_j + e'_i \tag{3}$$

where Y is the response, X_i and X_j are the variables, β is the regression coefficient, k is the number of variables tested and optimized in this experiment, and e is the random error. A p-value less than 0.05 was reported as significant.

Analytical Work

The chemical oxygen demand (COD), pH, biochemical oxygen demand (BOD_5), electrical conductivity (EC), total suspended solids (TSS), ammoniacal nitrogen (NH_3-N), and heavy metals (magnesium (Mg^{2+}), calcium (Ca^{2+}), iron (Fe^{2+}), zinc (Zn^{2+}), copper (Cu^{2+}), chromium (Cr^{2+}), cadmium (Cd^{2+}), lead (Pb^{2+}), arsenic (As^{3+}), cobalt (Co^{2+}), and manganese (Mn^{2+})) were tested before and after each aeration run. The level of BOD_5 was estimated utilizing Method 5210B. The DO was tested utilizing a DO meter (model 1000, YSI Inc., Greene County, OH, USA). COD concentration was tested utilizing the closed reflux colorimetric method (5220B—DR2500 HACH, Loveland, CO, USA). Color was determined using the DR 2800 HACH spectrophotometer at 455 nm wavelength.

A portable digital pH/mV meter (model inoLab pH 720, WTW, Weilheim, Germany) was used to measure the pH and EC. TSS was determined utilizing method 2540D, dried at 103–105 °C, which included the following procedure: preparation of filter disc, selection of filter type and sample volume, analysis of samples, and calculation of Equation (4) (APHA 2012) [31].

$$\text{TSS mg/L} = [(A - B) \text{ mg}/(V) \text{ mL}, \tag{4}$$

where A is the weight of filter-dried residue (mg), B is the weight of the filter, and V is the sample volume.

NH_3-N level was determined using the phenate method (4500-NH_3 F) utilizing a DR2500 spectrophotometer at 640 nm. Heavy metals were measured using atomic absorption spectroscopy (Unicam 929 AA Spectrophotometer, UNICO, Franksville, WI, USA). All physico-chemical parameters and heavy metals were measured according to standard methods for examining water and wastewater [32]. Different leachate/POME ratios (1:0, 0.9:0.1, 0.7:0.3, and 0.5:0.50) were prepared in 1000-mL bottles to investigate the removal efficiency of targeted parameters.

3. Results and Discussion

3.1. Effects of Aeration Time Variation on the Removal Efficiency during the Aeration Process of Leachate Treatment

The maximum removal efficiencies in only leachate aeration for COD, TSS, color, and NH_3-N reached 44.12%, 43.5%, 55%, and 97%. Figure 1 shows the removal efficiency influenced by the time of reaction for the targeted parameters. The removal efficiencies of both COD and TSS increased slightly in the same way. However, the increased color removal efficiency was characterized by fluctuation, as shown on Figure 1, while the removal efficiency of NH_3-N increased sharply in the first week of aeration, reaching 93% on the seventh day; it did not demonstrate a further marked increase in removal efficiency, reaching 97% by the 24th day. This is in line with many studies reporting that NH_3-N can be removed under the effect of the gas stripping process during aeration [21], while the removal is significantly improved with aeration time increasing up to a specific point [33] due to the reaction of NH_3 with water, as shown in Equation (5).

$$NH_3 + H_2O \leftrightarrow NH_4{}^+ + OH^-. \tag{5}$$

From Equation (5), increasing the pH due to the formation of OH^- will increase the concentration of NH_3. The increase in pH enhances the ammonia stripping during aeration [21] and enhances the removal of ammonia, after which, owing to the recarbonation of lime in leachate, the pH begins to decline by absorbing CO_2 from the ambient air [34].

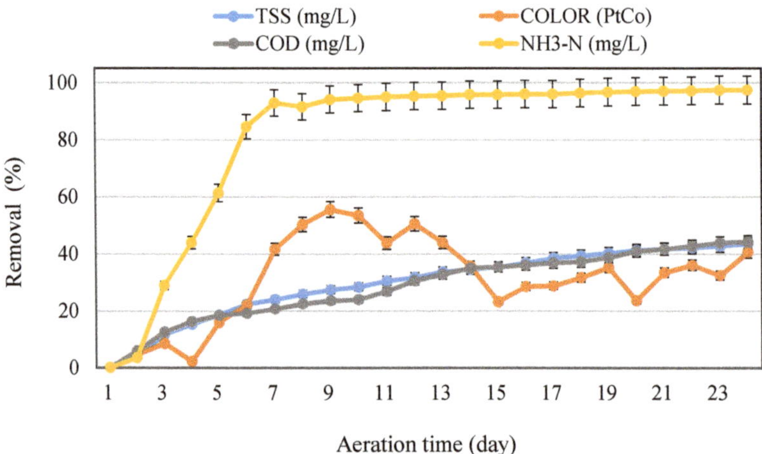

Figure 1. Effects of reaction time variation on COD, TSS, color, and NH_3-N removal efficiency (with natural pH, leachate only, aeration 20 L/min).

3.2. Effects of Reaction Time Variation on the Removal Efficiency during the Aeration Time of Leachate/POME Treatment (Ratio 900 mL Leachate/100 mL POME)

In this stage, the effect of the reaction time on leachate with a ratio of 900 mL leachate/100 mL POME was investigated during the aeration time. The optimum reaction time was reached on the 24th day of aeration, and the maximum removal efficiencies for COD, TSS, color, and NH_3-N reached 91%, 54%, 50%, and 98%, respectively. All targeted parameters increased as the aeration time increased (Figure 2).

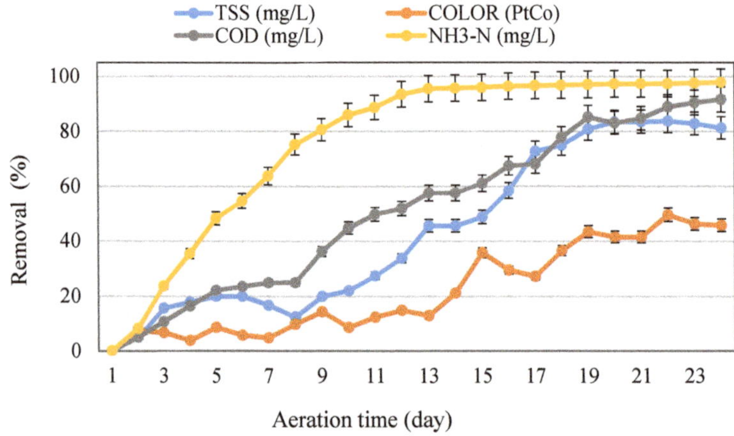

Figure 2. Effects of reaction time variation on COD, TSS, color, and NH_3-N removal efficiency (with natural pH, leachate/POME (900 mL leachate/100 mL POME), aeration power 20 L/min).

3.3. Effects of Reaction Time Variation on the Removal Efficiency during the Aeration Time of Leachate/POME Treatment (Ratio 700 mL Leachate/300 mL POME)

In this stage, the effects of reaction time variation on the removal efficiency during the aeration time of leachate/POME treatment with a constant ratio (700 mL leachate/300 mL POME) were investigated for the targeted parameters under the same condition of natural pH and an aeration power of 20 mL/min. The results showed an increase in the removal efficiency of all targeted parameters. The maximum removal of NH_3-N, COD, TSS, and color was 96%, 89%, 53%, and 41%, respectively. As shown in Figure 3, there was a higher removal efficiency for NH_3-N than COD, but color showed the lowest removal efficiency.

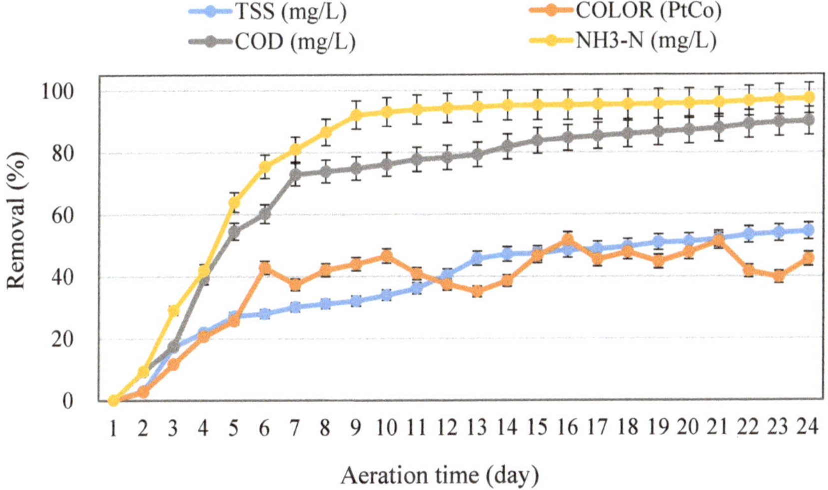

Figure 3. Effects of reaction time variation on COD, TSS, color, and NH_3-N removal efficiency (with natural pH, leachate/POME (700 mL leachate/300 mL POME), aeration power 20L/min).

3.4. Effects of Reaction Time Variation on the Removal Efficiency during the Aeration Time of Leachate/POME Treatment (Ratio 500 mL Leachate/500 mL POME)

As shown in Figure 4, the ratio between leachate and POME was 500 mL/500 mL. The effect of reaction time variation on the removal efficiency during the aeration time for targeted parameters was investigated with the same conditions for the other ratios (natural pH, aeration power 20 L/min). The maximum removal efficiencies for COD, TSS, color, and NH_3-N were 89%, 21%, 42%, and 94%. The removal efficiency of NH_3-N increased sharply in the first 10 days of aeration, reaching 90%, and it did not demonstrate a further marked increase in removal efficiency, reaching 94% by the 24th day.

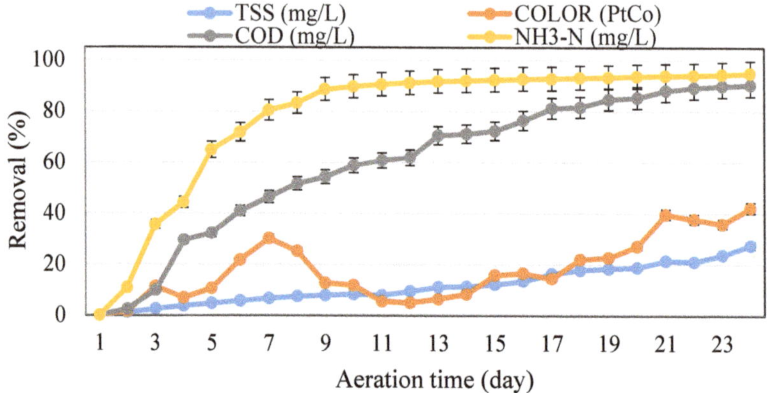

Figure 4. Effects of reaction time variation on COD, TSS, color, and NH$_3$-N removal efficiency (with natural pH, leachate/POME (500 mL leachate/500 mL POME), aeration power 20 L/min).

3.5. *Effects of the Leachate/POME Mixing Ratio on the Leachate Aeration Process*

The treatment of leachate was implemented using POME in four ratios (leachate only, 900 mL leachate/100 mL POME, 700 mL leachate/300 mL POME, and 500 mL leachate/500 mL POME) with a natural pH and an aeration power of 20 L/min. Accordingly, a limited removal efficiency of COD (43%) was found when treating leachate without any POME dosages. As shown in Figure 5, adding 100 mL, 300 mL, and 500 mL of POME improved the removal efficiency of COD (COD removal of 85%, 88%, and 88%, respectively).

The results of the removal efficiencies for the targeted parameters using several dosages (leachate only, 900 mL leachate/100 mL POME, 700 mL leachate/300 mL POME, and 500 mL leachate/500 mL POME) are illustrated in Figure 5. The maximum removal efficiency was NH$_3$-N, reaching 97% during the aeration process for leachate only and 900 mL leachate/100 mL POME, while the maximum removal for COD reached 88% (700 mL leachate/300 mL POME and 500 mL leachate/500 mL POME ratios). The lowest removal was 51% for TSS, followed by 52% for color for the 700 mL leachate/300 mL POME ratio.

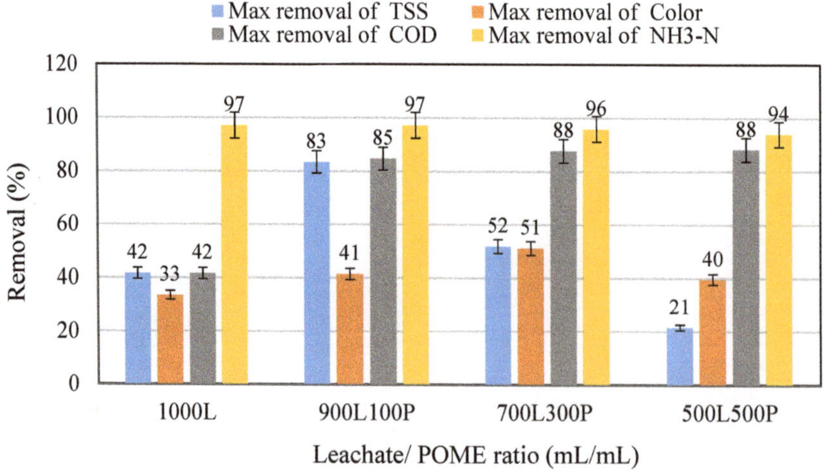

Figure 5. Effect of the leachate/POME ratios on COD, TSS, color, and NH$_3$-N removal efficiency after 24 days of mixing aeration (with natural pH, aeration power 20 L/min).

3.6. Analysis of Variance

One of the key objectives of the RSM is to calculate the best value for the control variables that can maximize or minimize a response over a particular area of concern. A good-fitting model is defined to provide a proper representation of the mean response to achieve the optimum value [34]. A total of 13 experiments with different dosages (leachate/POME) and reaction times at room temperature (28 °C) were performed using central composite design (CCD); the outcomes were analyzed using analysis of variance (ANOVA), as shown in Table 3. The POME dosage used to treat stabilized leachate was evaluated in terms of its effectiveness in the removal of COD, TSS, color, and NH_3-N.

The coefficient of determination (R^2) for COD, TSS, color, and NH_3-N was 0.9927, 0.8218, 0.9854, and 0.991, respectively. This mean that the models are good in Equations (6)–(9) due to the coefficient of determination being high and close to 1 (Table 4). Prior to the data analysis, the assumption of normality should be tested. The assumption of normality showed that the data roughly fit a bell-shaped curve for all responses, as presented in Figure 6.

$$\text{COD removal} = +80.18 + 25.81 \times A + 8.60 \times B - 19.97 \times A^2 - 4.00 \times B^2 - 0.98 \times A \times B; \tag{6}$$

$$\text{TSS removal} = +49.93 + 4.32 \times A + 15.76 \times B - 6.39 \times A^2 - 0.089 \times B^2 + 1.02 \times A \times B; \tag{7}$$

$$\text{Color removal} = +45.74 + 1.35 \times A + 16.14 \times B - 17.50 \times A^2 - 7.62 \times B^2 + 1.20 \times A \times B; \tag{8}$$

$$\text{NH}_3\text{-N removal} = +97.24 - 3.05 \times A + 6.72 \times B - 1.06 \times A^2 - 5.17 \times B^2 + 1.92 \times A \times B. \tag{9}$$

Table 3. Response value for different experimental conditions (POME dosages and aeration time).

Run	Factor A: POME (mL)	Factor B: Aeration Time (day)	COD Removal %			Color Removal %			TSS Removal %			NH$_3$-N Removal %		
			Actual	Predicted	Residual	Actual	Predicted	Residual	Actual	Predicted	Residual	Actual	Predicted	Residual
1	150	21	84.5	84.79	−0.29	51.89	55.1	−3.21	79	65.53	13.47	94	94.26	−0.26
2	150	14	78.81	80.18	−1.37	47.58	46.1	1.48	53.2	53.57	−0.37	93	92.83	0.17
3	150	14	78.31	80.18	−1.87	44.73	46.1	−1.37	51.3	53.57	−2.27	93	92.83	0.17
4	150	7	65.8	67.58	−1.78	24.21	21.37	2.84	25.6	36.92	−11.32	81	80.6	0.4
5	150	14	78.18	80.18	−2	47.69	46.1	1.59	54	53.57	0.43	93	92.83	0.17
6	300	14	84.9	86.02	−1.12	32.9	30.98	1.92	44	47.46	−3.46	90	88.93	1.07
7	0	14	33.46	34.4	−0.94	23.44	25.73	−2.29	36	30.38	5.62	94	94.93	−0.93
8	0	7	21.66	20.82	0.84	2.92	3.29	−0.37	18.46	15.07	3.39	85	84.7	0.3
9	300	7	75.32	74.39	0.93	1.5	3.97	−2.47	37.4	29.47	7.93	74	74.7	−0.7
10	0	21	40.08	39.98	0.099	35.1	32.44	2.66	32	41	−9	95	94.37	0.63
11	300	21	89.83	89.64	0.19	42.81	42.26	0.55	56.3	60.76	−4.46	92	92.37	−0.37
12	150	14	84.46	80.18	4.28	44.71	46.1	−1.39	54.4	53.57	0.83	93	92.83	0.17
13	150	14	83.22	80.18	3.04	46.15	46.1	0.052	52.8	53.57	−0.77	92	92.83	−0.83

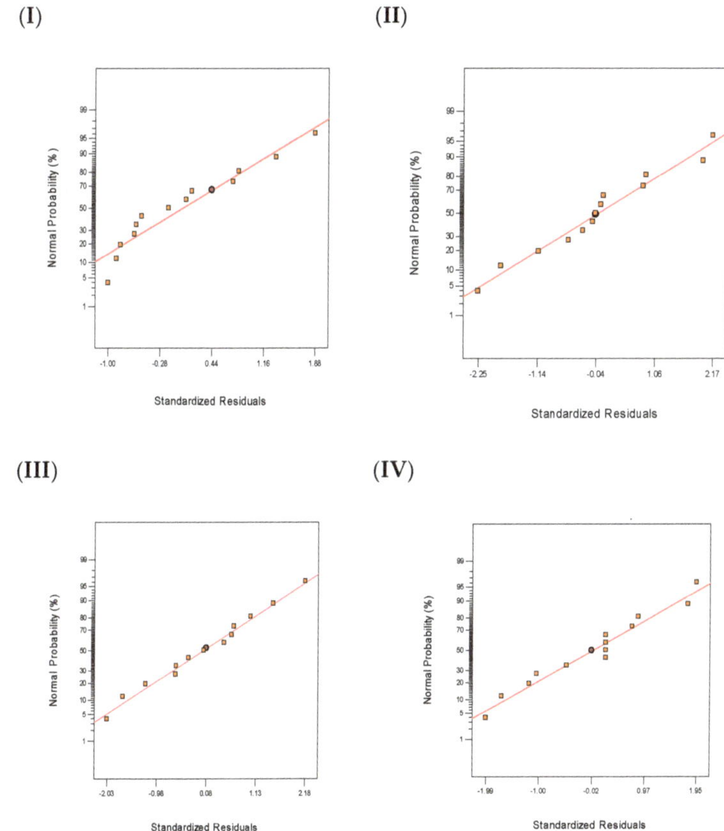

Figure 6. Normal probability plots for (**I**) COD, (**II**) TSS, (**III**) color, and (**IV**) NH$_3$-N removal.

The two variables, POME dosage and aeration time, displayed an important impact (p-value < 0.05) for the linear and quadratic models on the targeted factors of COD, color, and NH$_3$-N, as shown in Table 4 for the ANOVA results. On the other hand, a higher-order model like a third-order polynomial model or more complicated model can be used to improve the model for TSS removal.

Table 4. ANOVA for COD, TSS, color, and NH$_3$-N removal. A: POME dosage, B: aeration time.

	Source	Sum of Squares	DF	Mean Square	F Value	Prob > F
	Model	5981.97	5	1196.39	190.7	<0.0001
	A	3996.42	1	3996.42	637	<0.0001
	B	444.28	1	444.28	70.81	<0.0001
COD	A^2	1101.81	1	1101.81	175.62	<0.0001
Removal	B^2	44.26	1	44.26	7.06	0.0326
(%)	AB	3.82	1	3.82	0.61	0.4607
	Residual	43.92	7	6.27		
	Lack of Fit	7.85	3	2.62	0.29	0.8314
	Pure Error	36.07	4	9.02		
	Cor Total	6025.89	12			
	SD: 2.50, R^2: 0.9927, mean: 69.12, CV:3.62, Pred R^2: 0.9814, Adeq Precision: 40.447.					

Table 4. Cont.

	Source	Sum of Squares	DF	Mean Square	F Value	Prob > F
TSS Removal (%)	Model	2468.26	5	493.65	6.46	0.0148
	A	437.59	1	437.59	5.72	0.048
	B	1228.08	1	1228.08	16.06	0.0051
	A^2	592.46	1	592.46	7.75	0.0272
	B^2	15.2	1	15.2	0.2	0.6691
	AB	7.18	1	7.18	0.094	0.7681
	Residual	535.16	7	76.45		
	Lack of Fit	529.33	3	176.44	121.02	0.0002
	Pure Error	5.83	4	1.46		
	Cor Total	3003.43	12			

SD: 8.74, R^2: 0.8218, mean: 45.73, CV:19.12, Pred R^2: −0.7897, Adeq Precision: 8.495.

	Source	Sum of Squares	DF	Mean Square	F Value	Prob > F
Color Removal (%)	Model	3328.56	5	665.71	94.36	<0.0001
	A	41.34	1	41.34	5.86	0.046
	B	1705.89	1	1705.89	241.8	<0.0001
	A^2	869.58	1	869.58	123.26	<0.0001
	B^2	170.8	1	170.8	24.21	0.0017
	AB	20.84	1	20.84	2.95	0.1294
	Residual	49.39	7	7.06		
	Lack of Fit	40.88	3	13.63	6.41	0.0523
	Pure Error	8.5	4	2.13		
	Cor Total	3377.94	12			

SD: 2.66, R^2: 0.9854, mean: 34.28, CV: 7.75, Pred R^2: 0.8733, Adeq Precision: 28.712.

	Source	Sum of Squares	DF	Mean Square	F Value	Prob > F
NH_3-N Removal (%)	Model	458.76	5	91.75	154.36	<0.0001
	A	54	1	54	90.85	<0.0001
	B	280.17	1	280.17	471.33	<0.0001
	A^2	2.22	1	2.22	3.73	0.0446
	B^2	80.43	1	80.43	135.32	<0.0001
	AB	16	1	16	26.92	0.0013
	Residual	4.16	7	0.59		
	Lack of Fit	3.36	3	1.12	5.6	0.0647
	Pure Error	0.8	4	0.2		
	Cor Total	462.92	12			

SD: 0.77, R^2: 0.991, mean: 89.92, CV: 0.86, Pred R^2: 0.9238, Adeq Precision: 38.623.

Abbreviation. DF: degrees of freedom; Cor: corrected; CV: coefficient of variation; Pred: predicted; Adeq: adequate.

In addition, the p-values for the interaction effect were 0.4607, 0.7681, and 0.1294 for the removal efficiency of COD, TSS, and color. In other words, the interaction effect between POME dosage and aeration time was insignificant (p-value > 0.05) for the removal efficiency of COD, TSS, and color. That implies that the two factors function independently. Conversely, the p-value (0.0013) for the interaction effect was significant for NH_3-N removal efficiency. Figure 7 shows the interaction between the two (variables) factors (POME dosage and aeration time) and their behaviors in terms of removal of the targeted parameters (COD, TSS, color, and NH_3-N). The p-values for lack of fit were 0.8314, 0.0523, and 0.0647, which indicates that the lack of fit was insignificant (p-value > 0.05), which means that the model is appropriate for the removal efficiency of COD, color, and NH_3-N, while the p-value for lack of fit was significant (p-value < 0.05) for TSS removal efficiency, indicating the model is not appropriate.

The impact of POME dosage and aeration time on the selected responses is illustrated in Figure 8. The descriptions for the behavior of each response for POME dosage and aeration time are shown as the surface of a three-dimensional plot for the maximization of the four targeted responses (COD, TSS, color, and NH$_3$-N) (Figure 8). All response plots demonstrate clear peaks, suggesting that the maximum area of impact is well known with the selected boundaries of the POME dosage and aeration time.

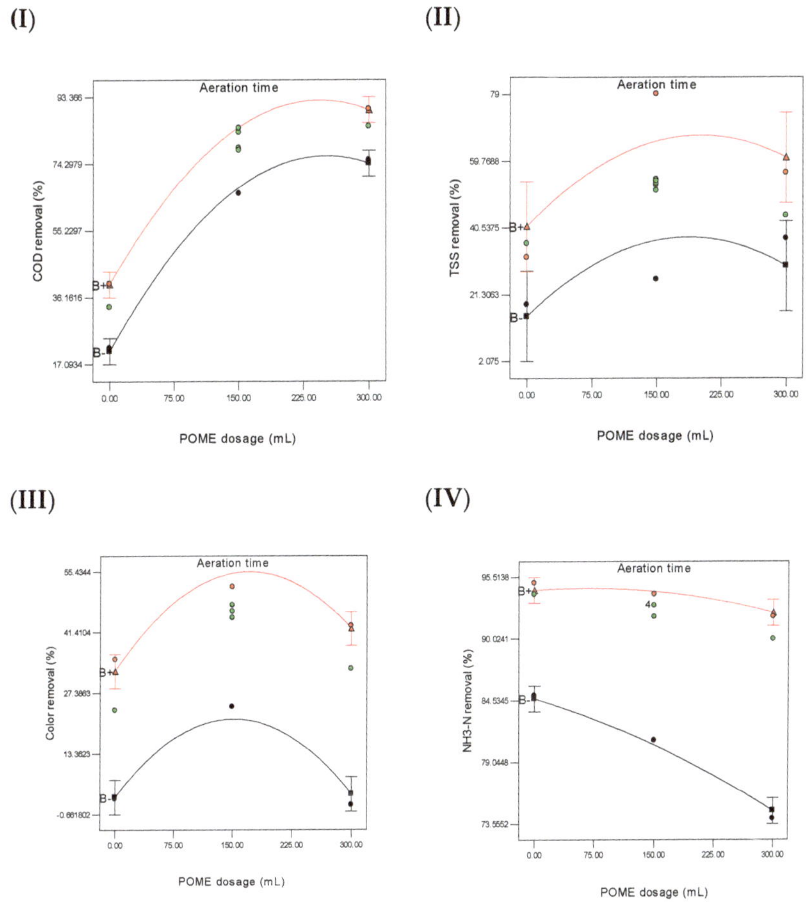

Figure 7. The impact of the POME dose combination and the aeration time for the removal of (**I**) COD, (**II**) TSS, (**III**) color, and (**IV**) NH$_3$-N.

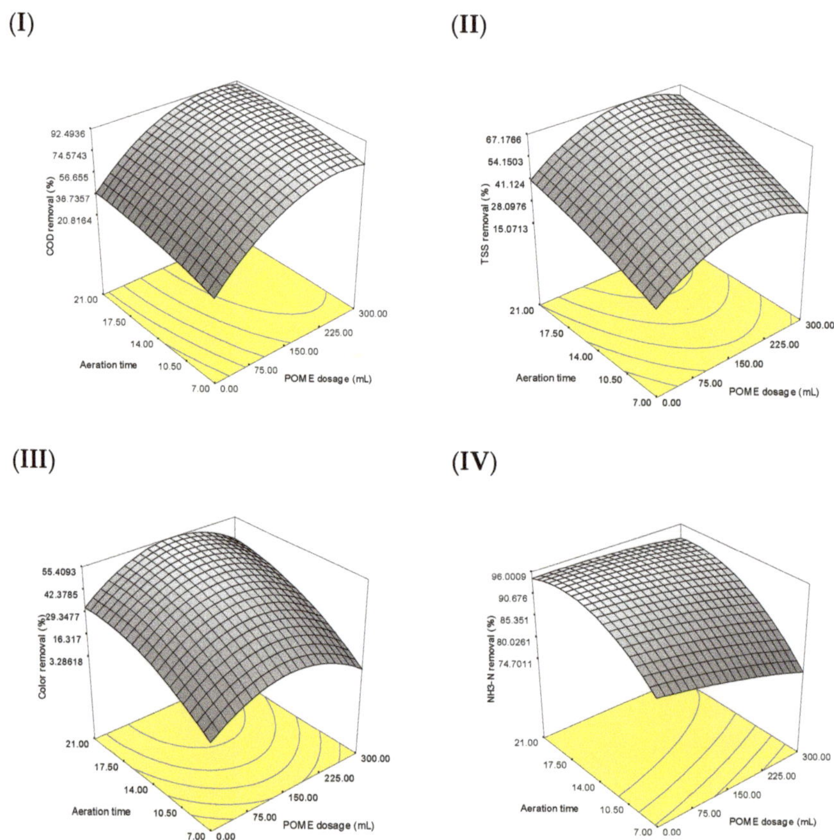

Figure 8. Response surface plot for (**I**) COD, (**II**) TSS, (**III**) color, and (**IV**) NH$_3$-N removal.

3.7. Optimization of Leachate Treatment Using POME

Design-Expert 6.0.7 software offers strong tools for setting up an optimal experiment of the treatment process to identify the optimum value of removal efficiency for COD, TSS, color, and NH$_3$-N. In accordance with the approach of software optimization, the required target was within the range for each experimental condition (POME dosage and aeration time). To obtain the highest output, the responses (COD, TSS, color, and NH$_3$-N) were described as a maximum value. The program incorporates individual desirability into a single number and then searches on the basis of the response target to optimize this feature. The optimum conditions and respective percentage removal efficiencies were established, and the COD (89.83%), TSS (66.7%), color (91.7%), and NH$_3$-N (94%) removal results are illustrated in Table 5 The desirability function for these optimum conditions was recorded as 0.935. Additional experiments under optimal conditions were performed to verify agreement with the outcome experiments from models and the experiments. The results from the laboratory experiment were 87.51%, 65.62%, 53.10%, and 91.8% for the removal of COD, TSS, color, and NH$_3$-N, respectively (Table 5). There was close agreement between the removal efficiencies for all response parameters gained from the experiments and those estimated by models. These results are more efficient than the result of Banch et al. [8], who conducted tests for the same parameters except for color. Tatsi et al. [35] achieved a color removal of about 100% for partially stabilized leachate. However, the reported residuals for COD, TSS, color, and NH$_3$-N were recorded as 430 mg/L, 1620 mg/L, 1780 Pt-Co, and 66 mg/L, respectively, which are still higher than the effluent discharge limits.

Table 5. Optimal response results from the model prediction and laboratory.

POME	Aeration Time	COD Removal (%)	TSS Removal (%)	Color Removal (%)	NH$_3$-N Removal (%)	Desirability
188.38	21.00	89.83	67.10	55.19	93.95	0.935
Lab experiment		87.15	65.54	52.78	91.75	

3.8. Heavy Metal Analysis

The removal efficiency of targeted heavy metals from stabilized leachate was evaluated under the achieved optimum experimental conditions, and the results are given in Table 6. The removal efficiency for the targeted heavy metals ranged between 99.00% (Cd^{2+}) and 6.85% (As^{3+}). The order of residuals of heavy metals in leachate was Fe^{2+} > Mn^{2+} > As^{3+} > Zn^{2+} > Cr^{2+} > Cu^{2+} > Co^{2+} > Pb$^+$ > Cd^{2+} from highest to lowest concentration. The heavy metal residuals were 1856, 31.72, 16.15, 10.76, 2.80, 1.28, 0.53, 0.28, and 0.23 µg/L for Fe^{2+}, Mn^{2+}, As^{3+}, Zn^{2+}, Cr^{2+}, Cu^{2+}, Co^{2+}, Pb$^+$, and Cd^{2+}, respectively, while the removal efficiency was 90.73%, 34.75%, 6.85%, 96.16%, 93.78%, 96.95%, 95.24%, 93.30%, and 99.00% respectively. Only the concentration of Mn^{2+} existed out of the limits, while the other targeted heavy metals were within the limits. The removal of heavy metals is attributed to the high level of suspended solids and metal complexes in POME which may act as a natural coagulant. The suspended solids and metal complexes improve the charged exchange and accelerated the adsorption and deposition of dissolved heavy metals in wastewater [36–40]. This study showed more efficient removal than Banch et al. [8] for Fe^{2+}, As^{3+}, Zn^{2+}, Cr^{2+}, Cu^{2+}, Co^{2+}, and Cd^{2+}.

Table 6. Effect of the POME dosage and aeration time on heavy metal removal (POME dosage 188.32, aeration time 21 days).

Heavy Metals	Initial Concentration in Leachate	Residual after Treatment Process	Removal (%)
Fe^{2+} (µg/L)	20.04 ± 7.11	1856 ± 0.57	90.73
Zn^{2+} (µg/L)	280.00 ± 19.63	10.76 ± 2.10	96.16
Cu^{2+} (µg/L)	41.91 ± 20.19	1.28 ± 0.64	96.95
Cr^{2+} (µg/L)	45.11 ± 12.81	2.80 ± 1.01	93.78
Cd^{2+} (µg/L)	22.62 ± 3.51	0.23 ± 0.20	99.00
Pb$^+$ (µg/L)	4.18 ± 2.91	0.28 ± 0.12	93.30
As^{3+} (µg/L)	17.34 ± 4.29	16.15 ± 2.37	6.85
Co^{2+} (µg/L)	11.05 ± 5.54	0.53 ± 0.39	95.24
Mn^{2+} (µg/L)	48.61 ± 14.99	31.72 ± 7.16	34.75

4. Conclusions

The current work evaluated the use of POME as an agro-industrial waste for landfill leachate treatment. Biological treatment using an aeration process for different mixing ratios between leachate and POME was performed. The optimization of the removal efficiencies for COD, TSS, color, and NH$_3$-N for the aeration treatment process of old leachate was investigated in this study. The optimum operational conditions were obtained at 188.32 mL of POME added to 1 L of leachate for 21 days of aeration time. The respective percentage removal efficiencies were 89.83%, 66.7%, 91.7%, and 94% for COD, TSS, color, and NH$_3$-N. Moreover, the treatment process reported an efficient removal of some heavy metals from landfill leachate. The results revealed that POME-based agricultural waste can be effectively used for organic removal from stabilized leachate.

Author Contributions: Formal analysis, T.J.H.B.; writing—original draft preparation, T.J.H.B.; writing—review and editing, M.M.H., S.S.A.A., A.F.M.A., and M.H.; supervision, M.M.H. and S.S.A.A.; funding acquisition, M.M.H. All authors read and agreed to the published version of the manuscript.

Funding: This research was partly funded by UKM research grants (DIP-2019-001 and MI-2020-005).

Acknowledgments: Marlia M. Hanafiah was supported by the Ministry of Education Malaysia (FRGS/1/2018/WAB05/UKM/02/2) and a UKM research grant (DIP-2019-001).

Conflicts of Interest: The authors declare no conflict of interest.

References

1. Gu, Z.; Chen, W.; Li, Q.; Zhang, A. Treatment of semi-aerobic aged-refuse biofilter effluent from treating landfill leachate with the Fenton method. *Process. Saf. Environ. Prot.* **2020**, *133*, 32–40. [CrossRef]
2. Guvenc, S.Y. Optimization of COD removal from leachate nanofiltration concentrate using H_2O_2/Fe^{+2}/heat-activated persulfate oxidation process. *Process. Saf. Environ. Prot.* **2019**, *126*, 7–17. [CrossRef]
3. Renou, S.; Givaudan, J.; Poulain, S.; Dirassouyan, F.; Moulin, P. Landfill leachate treatment: Review and opportunity. *J. Hazard. Mater.* **2008**, *150*, 468–493. [CrossRef] [PubMed]
4. Shalini, S.S.; Joseph, K.J. Nitrogen management in landfill leachate: Application of SHARON, ANAMMOX and combined SHARON–ANAMMOX process. *Waste Manag.* **2012**, *32*, 2385–2400. [CrossRef] [PubMed]
5. Abdulhasan, M.J.; Hanafiah, M.M.; Satchet, M.S.; Abdulaali, H.S.; Toriman, M.E.; AlRaad, A.A. Combining GIS, fuzzy logic and AHP models for solid waste disposal site selection in Nasiriyah, Iraq. *Appl. Ecol. Environ. Res.* **2019**, *17*, 6701–6722. [CrossRef]
6. Adhikari, B.; Dahal, K.R.; Khanal, S.N. A review of factors affecting the composition of municipal solid waste landfill leachate. *Int. J. Eng. Sci. Innov. Technol.* **2014**, *3*, 272–281.
7. Hermosilla, D.; Cortijo, M.; Huang, C. Optimizing the treatment of landfill leachate by conventional Fenton and photo-Fenton processes. *Sci. Total Environ.* **2009**, *407*, 3473–3481. [CrossRef]
8. Banch, T.J.; Hanafiah, M.M.; Alkarkhi, A.F.; Salem, A.M. Factorial design and optimization of landfill leachate treatment using tannin-based natural coagulant. *Polymers* **2019**, *11*, 1349. [CrossRef]
9. Banch, T.J.; Hanafiah, M.M.; Alkarkhi, A.F.; Salem, A.M. Statistical evaluation of landfill leachate system and its impact on groundwater and surface water in Malaysia. *Sains Malays.* **2019**, *48*, 2391–2403. [CrossRef]
10. Banch, T.J.; Hanafiah, M.M.; Alkarkhi, A.F.; Salem, A.M.; Nizam, N.U. Evaluation of different treatment processes for landfill leachate using low-cost agro-industrial materials. *Processes* **2020**, *8*, 111. [CrossRef]
11. Maslahati Roudi, A.; Chelliapan, S.; Wan Mohtar, W.; Kamyab, H. Prediction and optimization of the fenton process for the treatment of landfill leachate using an artificial neural network. *Water* **2018**, *10*, 595. [CrossRef]
12. Amr, S.S.A.; Aziz, H.A.; Adlan, M.N.J. Optimization of stabilized leachate treatment using ozone/persulfate in the advanced oxidation process. *Waste Manag.* **2013**, *33*, 1434–1441. [CrossRef] [PubMed]
13. Rada, E.; Istrate, I.; Ragazzi, M.; Andreottola, G.; Torretta, V. Analysis of electro-oxidation suitability for landfill leachate treatment through an experimental study. *Sustainability* **2013**, *5*, 3960–3975. [CrossRef]
14. Bashir, M.J.; Aziz, H.A.; Amr, S.S.A.; Sethupathi, S.A.P.; Ng, C.A.; Lim, J. The competency of various applied strategies in treating tropical municipal landfill leachate. *Desalin. Water Treat.* **2015**, *54*, 2382–2395. [CrossRef]
15. Mojiri, A. Review on membrane bioreactor, ion exchange and adsorption methods for landfill leachate treatment. *Aust. J. Basic Appl. Sci.* **2011**, *5*, 1365–1370.
16. Mussa, Z.H.; Othman, M.R.; Abdullah, M.P. Electrochemical oxidation of landfill leachate: Investigation of operational parameters and kinetics using graphite-PVC composite electrode as anode. *J. Braz. Chem. Soc.* **2015**, *26*, 939–948. [CrossRef]
17. Aziz, N.I.H.A.; Hanafiah, M.M. Life cycle analysis of biogas production from anaerobic digestion of palm oil mill effluent. *Renew. Energy* **2020**, *145*, 847–857. [CrossRef]
18. Aziz, N.I.H.A.; Hanafiah, M.M.; Ali, M.Y. Sustainable biogas production from agrowaste and effluents—A promising step for small-scale industry income. *Renew. Energy* **2019**, *132*, 363–369. [CrossRef]
19. Ma, W.; Han, Y.; Ma, W.; Han, H.; Zhu, H.; Xu, C.; Li, K.; Wang, D. Enhanced nitrogen removal from coal gasification wastewater by simultaneous nitrification and denitrification (SND) in an oxygen-limited aeration sequencing batch biofilm reactor. *Bioresour. Technol.* **2017**, *244*, 84–91. [CrossRef]
20. Aziz, N.I.H.A.; Hanafiah, M.M.; Gheewala, S. A review on life cycle assessment of biogas production: Challenges and future perspectives in Malaysia. *Biomass Bioenergy* **2019**, *122*, 361–374. [CrossRef]
21. Hao, Y.J.; Ji, M.; Chen, Y.X.; Wu, W.X.; Hao, Y.J.; Zhang, S.G.; Liu, H.Q. The pathway of in-situ ammonium removal from aerated municipal solid waste bioreactor: Nitrification/denitrification or air stripping? *Waste Manag. Res.* **2010**, *28*, 1057–1064. [CrossRef] [PubMed]

22. Daud, Z.; Awang, H. Integrated Leachate Treatment Technology. In *Waste Management: Concepts, Methodologies, Tools, and Applications*; IGI Global: Hershey, PA, USA, 2020; pp. 204–220.
23. Jagaba, A.H.; Latiff, A.A.; Latiff, A.; Umaru, I.; Abubakar, S.; Lawal, I. Treatment of Palm Oil Mill Effluent (POME) by Coagulation-Flocculation using Different Natural and Chemical Coagulants: A Review. *IOSR J. Mech. Civ. Eng.* **2016**, *13*, 67–75.
24. Mohamed, A.F.; Yaacob, W.W.; Taha, M.R.; Samsudin, A.R. Groundwater and soil vulnerability in the Langat Basin Malaysia. *Eur. J. Sci. Res.* **2009**, *27*, 628–635.
25. Hanafiah, M.M.; Yussof, M.K.M.; Hasan, M.; AbdulHasan, M.J.; Toriman, M.E. Water quality assessment of Tekala River, Selangor, Malaysia. *Appl. Ecol. Environ. Res.* **2018**, *16*, 5157–5174. [CrossRef]
26. Thaldiri, N.H.; Hanafiah, M.M.; Halim, A.A. Effect of modified micro-sand, poly-aluminium chloride and cationic polymer on coagulation-flocculation process of landfill leachate. *Environ. Ecosyst. Sci.* **2017**, *1*, 17–19. [CrossRef]
27. Banch, T.J.H.; Hanafiah, M.M.; Amr, S.S.A.; Ashraf, M.A. Characterization of leachate from Ampar Tenang closed landfill site, Selangor, Malaysia. *AIP Conf. Proc.* **2019**, *2111*, 060009.
28. Yusoff, I.; Alias, Y.; Yusof, M.; Ashraf, M.A. Assessment of pollutants migration at Ampar Tenang landfill site, Selangor, Malaysia. *Sci. Asia* **2013**, *39*, 392–409. [CrossRef]
29. US Environmental Protection Agency. Effluent Limitations Guidelines, Pretreatment Standards and New Source Performance Standards for the Landfills Point Source Category. 2000. Available online: https://www.federalregister.gov/documents/2000/01/19/00--1037/effluent-limitations-guidelines-pretreatment-standards-and-new-sourceperformance-standards-for-the (accessed on 7 January 2020).
30. *Environmental Quality (Control of Pollution from Solid Waste Transfer Station and Landfill) Regulations Malaysia*; FAO: Rome, Italy, 2009.
31. APHA-American Public Health Association. *2540 Solids, 2540-D Total Suspended Solids Dried at 103–105 °C*; Alpha Analytical, Inc.: Holmes, PA, USA, 2012.
32. US Environmental Protection Agency. *Standard Methods for the Examination of Water and Wastewater*; American Public Health Association: Washington, DC, USA, 2005; Volume 2.
33. Feron, P. (Ed.) *Absorption-Based Post-Combustion Capture of Carbon Dioxide*; Woodhead Publishing: Cambridge, UK, 2016.
34. Bonmatı, A.; Flotats, X. Air stripping of ammonia from pig slurry: Characterisation and feasibility as a pre-or post-treatment to mesophilic anaerobic digestion. *Waste Manag.* **2003**, *23*, 261–272. [CrossRef]
35. Tatsi, A.A.; Zouboulis, A.I.; Matis, K.A.; Samaras, P. Coagulation–flocculation pretreatment of sanitary landfill leachates. *Chemosphere* **2003**, *53*, 737–744. [CrossRef]
36. Ashraf, M.A.; Balkhair, K.S.; Chowdhury, A.J.K.; Hanafiah, M.M. Treatment of Taman Beringin landfill leachate using the column technique. *Desalin. Water Treat.* **2019**, *149*, 370–387. [CrossRef]
37. Manikam, M.K.; Halim, A.A.; Hanafiah, M.M.; Krishnamoorthy, R.R. Removal of ammonia nitrogen, nitrate, phosphorus and COD from sewage wastewater using palm oil boiler ash composite adsorbent. *Desalin. Water Treat.* **2019**, *149*, 23–30. [CrossRef]
38. Hanafiah, M.M.; Zainuddin, M.F.; Nizam, N.U.M.; Rasool, A. Phytoremediation of Aluminium and Iron from Industrial Wastewater Using *Ipomoea aquatica* and *Centella asiatica*. *Appl. Sci.* **2020**, *10*, 3064. [CrossRef]
39. Nizam, N.U.M.; Hanafiah, M.M.; Noor, I.M.; Karim, H.I.A. Efficiency of five selected aquatic plants in phytoremediation of aquaculture wastewater. *Appl. Sci.* **2020**, *10*, 1–11.
40. Asman, N.S.; Halim, A.A.; Hanafiah, M.M.; Ariffin, F.D. Determination of rainwater quality from rainwater harvesting system at Ungku Omar College, UKM Bangi. *Sains Malays.* **2017**, *46*, 1211–1219. [CrossRef]

 *processes*

Article

Effect of Ionic Strength and Coexisting Ions on the Biosorption of Divalent Nickel by the Acorn Shell of the Oak *Quercus crassipes* Humb. & Bonpl.

Erick Aranda-García [1], Griselda Ma. Chávez-Camarillo [2] and Eliseo Cristiani-Urbina [1,*

[1] Departamento de Ingeniería Bioquímica, Escuela Nacional de Ciencias Biológicas, Instituto Politécnico Nacional, Ciudad de México 07738, Mexico; arand241@hotmail.com
[2] Departamento de Microbiología, Escuela Nacional de Ciencias Biológicas, Instituto Politécnico Nacional, Ciudad de México 11340, Mexico; gchsepi@gmail.com
* Correspondence: ecristiani@ipn.mx; Tel.: +52-55-57296000 (ext. 57835)

Received: 19 August 2020; Accepted: 28 September 2020; Published: 1 October 2020

Abstract: This study investigated the effect of ionic strength and background electrolytes on the biosorption of Ni^{2+} from aqueous solutions by the acorn shell of *Quercus crassipes* Humb. & Bonpl. (QCS). A NaCl ionic strength of 0.2 mM was established to have no effect on the Ni^{2+} biosorption and the biosorption capacity of the heavy metal decreased as the ionic strength increased from 2 to 2000 mM. The background electrolytes (KCl, $NaNO_3$, Na_2SO_4, $CaCl_2$, $MgSO_4$, and $MgCl_2$) had no adverse effects on the biosorption of Ni^{2+} at a concentration of 0.2 mM. However, at background electrolyte concentrations of 2 and 20 mM, divalent cations (Ca^{2+} and Mg^{2+}) had greater negative effects on the biosorption of Ni^{2+} compared to the monovalent cations (Na^+ and K^+). Additionally, the SO_4^{2-} and Cl^- anions affected the biosorption of Ni^{2+}. The fractional power, Elovich, and pseudo-second order models represented the kinetic processes of the biosorption of Ni^{2+} adequately. The results show that QCS can be a promising and low-cost biosorbent for removing Ni^{2+} ions from aqueous solutions containing various types of impurities with different concentrations.

Keywords: biosorbent; Ni^{2+}; background electrolytes; kinetic modeling

1. Introduction

Currently, environmental pollution by toxic heavy metals is one of the most alarming problems of modern society [1–3]. Since heavy metals are non-biodegradable and highly toxic, their presence in water resources poses a great risk to the balance of the natural environment and the health of living beings [4,5].

The divalent Ni^{2+} is one of the most toxic heavy metals found in wastewater discharges owing to various anthropogenic activities, such as the manufacture of metal alloys, stainless steel, super-alloys, accumulators, batteries, electrical and electronic products and components, pigments, paints, coins, and ceramics, mineral processing, steel casting, nickel mining and refining, metallurgy, electroplating, leather tanning, and porcelain enameling [3,6,7]. Notably, it is evident that Ni^{2+} is widely used in several industrial sectors, including transportation, construction, electronics, aeronautics, automotive, and telecommunications [8].

Exposure to high levels of Ni^{2+} causes a range of harmful effects on human health, such as endocrine disorders, gastrointestinal distress, allergies, headache, anemia, dizziness, chest tightness, pulmonary fibrosis, cyanosis, rapid breathing, and encephalopathy, as well as damage to the kidneys, central nervous system, and lungs [1,4,9–11]. Moreover, Ni^{2+} exhibits carcinogenic, embryotoxic, and teratogenic properties [9,10]. Therefore, to protect the public health from the harmful effects of

Ni^{2+}, the World Health Organization (WHO) established a reference value of 0.07 mg/L to control the concentration of nickel in drinking water [12].

The conventional methods used to remediate industrial wastewater contaminated with Ni^{2+}, such as chemical coagulation and precipitation, adsorption onto activated carbon, ion exchange, and various electrochemical and membrane technologies [8,13] have several disadvantages. These disadvantages include high cost, inefficient or ineffective treatment of wastewater with low Ni^{2+} concentrations, production of toxic chemical sludge that requires additional treatment, and/or they are highly sensitive to the operating parameters [9,14]. These disadvantages together with the increasing implementation of stricter environmental regulations have prompted the search for new treatment technologies [13]. Biosorption is a cost-effective, flexible, and efficient technology for removing heavy metals from aqueous solutions, which uses plant, animal, and microbial biomass or their derived products as biosorbents [15–17]. Agricultural and forestry residues and by-products, which are mainly composed of cellulose, hemicellulose, and lignin, are abundant in nature, renewable, economical, and environmental friendly. Additionally, they are highly efficient and effective for removing organic and inorganic contaminants from aqueous solutions via biosorption. Therefore, they are a viable option for bioremediation of industrial effluents contaminated with heavy metals [2,8,13,18,19].

Our previous studies established that the acorn shell of *Quercus crassipes* Humb. and Bonpl. (QCS) is a versatile and effective novel biosorbent for removing anionic and cationic heavy metals from aqueous solutions. QCS has a remarkable ability to remove hexavalent chromium (anionic heavy metal in aqueous solution) and to biosorb total chromium from aqueous solutions, both in batch [20,21] and continuous [22] systems.

Furthermore, so far, QCS is one of the best biosorbents reported for the biosorption of Ni^{2+} (cationic heavy metal) from aqueous solutions. Therefore, it was established that the QCS performance in the biosorption of Ni^{2+} ions is affected by the contact time, pH of the solution, initial Ni^{2+} concentration, and temperature. The optimal pH for the biosorption of Ni^{2+} by QCS is 8.0, whereas its point of zero charge is 5.4. The kinetic and equilibrium biosorption processes of Ni^{2+} are significantly represented using the pseudo-second order and Freundlich models, respectively. Moreover, it was established that the biosorption of Ni^{2+} by QCS is an endothermic process, non-spontaneous, and of chemical nature, in which the carboxyl, carbonyl, and hydroxyl functional groups play a major role in the removal of the heavy metal [9].

One of the critical parameters to be considered in the scaling up and large-scale application of biosorption processes is the presence of co-ions in the wastewater to be treated [20,23]. Therefore, it is important to note that most studies on biosorption of toxic heavy metals have been carried out using synthetic solutions that contain the metal of interest only. However, real industrial effluents are usually complex mixtures containing different types of background electrolytes, such as monovalent and divalent cations and anions at different concentrations [24]. The background electrolytes and their concentrations can affect the biosorption of the heavy metal of interest since they can: (1) compete with the heavy metal of interest for the available biosorption active sites, (2) decrease the specificity of the biosorbent by binding to sites to which the metal ion of interest does not bind, and/or (3) form chemical complexes or precipitates with the heavy metal of interest [20,23,24].

In spite of their great importance and relevance for the biosorption processes of toxic heavy metals, there is practically no information in the specialized literature about the effects of ionic strength and competing ions on the biosorptive removal of heavy metals from aqueous solutions [25]. This information is crucial for analyzing, interpreting, understanding and designing biosorption processes for heavy metal removal from aqueous solutions, meaning there is a clear need for investigation concerning the inhibitory effects of ionic strength and competing ions on the biosorption of the heavy metal of interest.

Therefore, the aims of the current investigation are to assess the influences of background cations, background anions, and NaCl ionic strength, on both the biosorption of Ni^{2+} ions onto QCS in aqueous solution, and the kinetic modeling of the Ni^{2+} biosorption process.

2. Materials and Methods

2.1. Biosorbent

The acorns of the oak *Quercus crassipes* Humb. and Bonpl. were collected from the town of El Durazno de Cuauhtémoc, located in the municipality of Jilotepec de Molina Enríquez, in the state of Mexico, Mexico. The acorns were washed under running water, rinsed with distilled deionized water, and then dried in an oven at 60 °C for 24 h. Thereafter, the shells were separated from the acorns and grounded in a Glen Creston® laboratory mill (Glen Creston, Ltd., London, England, UK). The resulting powder was sieved using American Society for Testing and Materials (ASTM) sieves (ASTM International, West Conshohocken, PA, USA), and the fraction with particle sizes ranging between 180–212 μm was used in all the experiments carried out in this study.

2.2. Stock and Test Solutions

Stock solutions of $NiSO_4$, KCl, $NaNO_3$, Na_2SO_4, $CaCl_2$, $MgSO_4$, and $MgCl_2$ with a concentration of 20 mM and 2000 mM NaCl were prepared by dissolving a precisely weighed amount of chemical compounds in 1 L distilled deionized water. All reagents were of analytical grade (JT Baker®, Monterrey, Mexico). The test solutions were prepared by diluting the stock solutions with distilled deionized water.

2.3. Biosorption Experiments

The kinetic experiments were carried out using batch systems to assess the influence of ionic strength and background electrolytes on the biosorption of Ni^{2+} from aqueous solutions by the acorn shell of *Quercus crassipes* Humb. & Bonpl. (QCS). The biosorption studies were performed in 500 mL Erlenmeyer flasks containing 110 mL Ni^{2+} solution at an initial concentration of 1.97 mM and a QCS biomass at a concentration of 1 g/L. The flasks were shaken at a constant speed of 120 rpm in a Cole-Parmer® linear shaking water bath (Cole-Parmer®, Vernon Hills, IL, USA) for 120 h at 25 ± 1 °C. The pH of each test solution was measured regularly over the course of the experiments and then adjusted to 8.0 ± 0.1 using 0.1 M NaOH and/or HCl solutions when necessary.

Ni^{2+} solutions containing some of the anions and cations that have been frequently found in industrial effluents were used to assess the influence of ionic strength and background electrolytes on the biosorption of Ni^{2+} by QCS [24]. The effect of ionic strength was tested using NaCl as the background electrolyte at concentrations ranging from 0.2 to 2000 mM. To assess the effect of background electrolytes, chemical compounds (KCl, $NaNO_3$, Na_2SO_4, $CaCl_2$, $MgSO_4$, and $MgCl_2$) consisting of monovalent and divalent anions and cations were used at concentrations of 0.2, 2, and 20 mM. Control experiments that only contained QCS biomass at a concentration of 1 g/L and Ni^{2+} solution at an initial concentration of 1.97 mM with no background electrolytes were conducted simultaneously.

Additionally, control experiments with no QCS biomass were performed using the same operating conditions of the Ni^{2+} biosorption experiments to determine whether there was loss of Ni^{2+} because of precipitation or adsorption onto the glass. Statistically, there were no significant differences between the Ni^{2+} concentrations in the control experiments with no QCS biomass at different experimental times. Thus, the decrease in the Ni^{2+} concentration observed in the experiments with QCS biomass was caused by QCS.

Samples were taken at different experimentation times, and then they were filtered through a Whatman® grade 42 filter paper (Whatman®, St. Louis, MI, USA). The collected filtrates were analyzed to determine their residual Ni^{2+} concentration.

The Ni^{2+} biosorption capacity (q_t, mmol/g) at any time t was calculated using Equation (1):

$$q_t = \frac{(C_i - C_t)}{C_b} \tag{1}$$

where C_i (mmol/L) is the initial concentration of Ni^{2+} at time t = 0 h, C_t (mmol/L) is the residual concentration of Ni^{2+} at time t = t, and C_b is the concentration of the QCS biosorbent (g/L).

The effect of background electrolytes and ionic strength on the biosorption of Ni^{2+} by QCS was quantitatively assessed using the global performance index of the Ni^{2+} biosorption (ξ, %), which was calculated using Equation (2) [26]:

$$\xi = 100 \frac{\left(\int_{t=0}^{t=t_f} q_t dt\right)_{problem} - \left(\int_{t=0}^{t=t_f} q_t dt\right)_{control}}{\left(\int_{t=0}^{t=t_f} q_t dt\right)_{control}} \tag{2}$$

where $(q_t dt)_{problem}$ and $(q_t dt)_{control}$ are the time courses of the Ni^{2+} biosorption capacity in the experiments carried out with (test experiments) and without (control experiments) background electrolytes, respectively, and t_f is the total contact time between the Ni^{2+} solution and the QCS biomass (120 h).

Integration of the $(q_t dt)_{problem}$ and $(q_t dt)_{control}$ functions from t = 0 to t = t was performed using Mathematica version 7.0 (Wolfram Research, Champaign, IL, USA). The biosorption of Ni^{2+} by QCS is not affected by the ionic strength or background electrolytes if $\xi = 0\%$. There is an improvement in the biosorption of Ni^{2+} compared to the control if $\xi > 0\%$, thus, the ionic strength or background electrolytes have a positive or synergistic effect on the metal biosorption. Finally, if $\xi < 0\%$, the biosorption of Ni^{2+} is decreased compared to the control, thus, the ionic strength or background electrolytes have a negative or antagonistic effect on the metal biosorption [26].

2.4. Determination of the Ni^{2+} Concentration

The Ni^{2+} concentration in the liquid phase was determined using the dimethylglyoxime method [27]. The absorbance of the chemical complex with a color ranging from red wine to brown that is formed during the reaction of Ni^{2+} ions with dimethylglyoxime was measured in a Thermo Scientific™ Evolution 201 spectrophotometer (Thermo Fisher Scientific, Whaltman, MA, USA) at a wavelength of 465 nm.

2.5. Biosorption Kinetic Modeling

The biosorption kinetics describes the rate of biosorption of the adsorbate and, therefore, controls the time required to reach dynamic equilibrium [28]. This study used the pseudo-first order, pseudo-second order, Elovich, intraparticle diffusion, and fractional power models to analyze the kinetics of the Ni^{2+} biosorption process.

The pseudo-first order and pseudo-second order models can be described using Equations (3) and (4), respectively [29]:

$$q_t = q_{e1} \left(1 - e^{-k_1 t}\right) \tag{3}$$

$$q_t = \frac{t}{\frac{1}{k_2 q_{e2}^2} + \frac{t}{q_{e2}}} \tag{4}$$

where q_{e1} and q_{e2} are the equilibrium biosorption capacities (mmol/g) predicted by the pseudo-first order and pseudo-second order models, respectively, q_t is the biosorption capacity (mmol/g) at time t = t (h), and k_1 (1/h) and k_2 (g/mmol·h) are the rate constants of the pseudo-first order and pseudo-second order models, respectively.

The Elovich kinetic model is given by Equation (5) [30]:

$$q_t = \frac{1}{B_e} \ln(1 + A_e B_e t) \tag{5}$$

where A_e (mmol/g·h) and B_e (g/mmol) are the initial biosorption rate and the desorption constant, respectively.

The fractional power kinetic model is given by Equation (6) [31]:

$$q_t = k_{fp}\, t^v \tag{6}$$

where k_{fp} (mmol/g) is the fractional power model constant and v (1/h) is the rate constant of the fractional power model. The product of these constants ($k_{fp} \cdot v$, mmol/g·h) is known as the specific biosorption rate at unit time, that is, when $t = 1$.

The intraparticle diffusion model can be described by Equation (7) [32]:

$$q_t = k_{id}\, t^{0.5} + c \tag{7}$$

where c (mmol/g) is the intercept of the model that is related to the thickness of the boundary layer and k_{id} (mmol/g·h$^{0.5}$) is the intraparticle diffusion rate constant.

2.6. Determination of the Parameters of the Kinetic Models and Statistical Analysis

All the experiments of the study were carried out at least twice and the results reported herein are the average values ± mean standard deviation. Statistical analysis of the Ni^{2+} biosorption data and estimation of the parameters of the tested kinetic models were performed using the GraphPad Prism version 8.0 (GraphPad Software, Inc., San Diego, CA, USA).

The analysis of variance (ANOVA) was performed using the Tukey's test for group comparisons to determine whether there were statistically significant differences in the Ni^{2+} biosorption data. A significance level (α) of 0.05 was used. Probability values (p) lower than α indicate that the evaluated groups differ significantly.

The kinetic models parameters were obtained via non-linear regression analysis of the experimental data. Various error functions, such as root mean squared error (RMSE), sum of squared error (SSE), Akaike information criterion (AIC), coefficient of determination (r^2), and 95% confidence intervals were used to determine the accuracy and adequacy of the tested kinetic models fit. Small RMSE, SSE, and AIC values, r^2 values close to one, and narrow 95% confidence intervals indicate that the models describe the experimental Ni^{2+} biosorption data more accurately [22].

3. Results and Discussion

3.1. Influence of Ionic Strength on the Biosorption of Ni^{2+} by Acorn Shell of Quercus Crassipes Humb. & Bonpl. (QCS)

The ionic strength of an aqueous solution is an environmental parameter that significantly affects the biosorption of heavy metal ions at the interface between the solid biosorbent and the liquid phase [24]. Thus, this study investigated its effect on the kinetics of biosorption of Ni^{2+} by QCS.

Figure 1 shows the variation in Ni^{2+} biosorption capacity as a function of the biosorption time at NaCl ionic strengths ranging from 0.2 to 2000 mM. The kinetic profile of the Ni^{2+} biosorption at an ionic strength of 0.2 mM is similar to that of the control experiment (0 mM NaCl). There were no significant differences between the Ni^{2+} biosorption capacities ($p > 0.05$) for the control experiment and at an ionic strength of 0.2 mM at different experimental times, which indicates that the Ni^{2+} biosorption is not affected by a NaCl ionic strength of 0.2 mM.

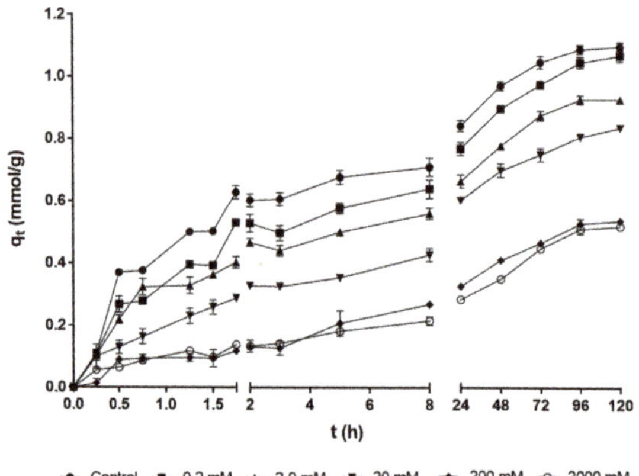

Figure 1. Effect of the NaCl ionic strength on the biosorption of Ni^{2+} by acorn shell of *Quercus crassipes* Humb. & Bonpl. (QCS).

In contrary, the Ni^{2+} biosorption capacity decreased gradually as the ionic strength increased from 0.2 to 200 mM ($p < 0.05$). However, the Ni^{2+} biosorption capacities were very similar and there were no significant differences between them ($p > 0.05$) at different biosorption times and at very high ionic strengths of 200 and 2000 mM. These results indicate that the biosorption of Ni^{2+} is negatively affected by NaCl ionic strengths equal to or greater than 2 mM.

The global performance indexes for the biosorption of Ni^{2+} at the different ionic strengths are summarized in Table 1.

Table 1. Effect of the NaCl ionic strength on the global performance index of the biosorption of Ni^{2+} by acorn shell of *Quercus crassipes* (QCS).

Ionic Strength (mM)	ξ (%)
0.2	−4
2	−17
20	−26
200	−55
2000	−58

The global performance index revealed that a NaCl ionic strength of 0.2 mM had a negligible negative effect on the biosorption of the metal ($ξ = -4\%$). However, as the ionic strength increased from 2 to 200 mM, the global performance index decreased from −17 to −55%, which indicates that the adverse effect of the ionic strength on the biosorption of Ni^{2+} increases with an increase in ionic strength ($p < 0.05$). Moreover, the global performance index for the biosorption of Ni^{2+} at NaCl ionic strengths of 200 and 2000 mM were similar, thus, the adverse effects on the biosorption of the metal were similar under these conditions ($p > 0.05$). These results are in agreement with the negative effects observed in the kinetic studies of the Ni^{2+} biosorption.

The decrease in Ni^{2+} biosorption with increasing NaCl ionic strengths from 0.2 to 2000 mM can be attributed to the fact that the aqueous solution contain more positively charged Na^+ ions, which competed with the Ni^{2+} cations for the available biosorption active sites on the QCS surface [24,33]. Moreover, changes in the ionic strength of a solution can make the reactive functional groups on the surface of the biosorbent less accessible to Ni^{2+} ions [20,34]. Furthermore, Na^+ ions can reduce the concentration of other electrostatically bound counterions that balance the negative charge of the biomass, thus, Na^+ ions affect the intraparticle ion concentration and the binding of other ions, such as the Ni^{2+} ions [33]. Additionally, it has been reported that high Cl^- ion concentrations (present in NaCl) favor the formation of the nickel chloride ($NiCl^-$) complex, thus decreasing the number of free Ni^{2+}

ions in the aqueous solution, which results in a lower number of interactions between Ni^{2+} ions and biosorption active sites and consequently, a decrease in the heavy metal biosorption capacity [35].

Additionally, it has been reported that the NaCl ionic strength negatively affects the biosorption of Ni^{2+} by grape stalks wastes [35], filamentous fungi such as *Rhizopus* sp., *Mucor* sp., and *Penicillium* sp. [36], and barley straw [37]. Moreover, the NaCl ionic strength decreased the sorption capacity of other heavy metals, such as Cu^{2+} by HNO_3-pretreated newspaper scraps, HNO_3-pretreated-maize spatha [38], and by an exopolysaccharide of *Wangia profunda* [39]; also Cd^{2+} by an exopolysaccharide of *Wangia profunda* [39] and by magnetic graphene oxide-supported sulfanilic acid [24], Pb^{2+} by *Sargassum filipendula* [40], and Cr^{6+} by lignin [28]. Likewise, the increase of NaCl ionic strength inhibited the biosorption of organic adsorbates, such as Methylene Blue and Rhodamine B dyes by *Phellinus igniarius* and *Fomes fomentarius* [41], and benzene and toluene by *Macrocystis pyrifera* [42].

Analysis of the kinetic model for biosorption study is crucial for understanding dynamics, mechanism, and reaction pathway of the biosorption process. It also helps in determining mass transfer, rate-controlling steps, and physicochemical interaction in the biosorption process. Additionally, knowledge about kinetics of metal biosorption is useful for determining optimum conditions for biosorption processes [43]. In the present work, the experimental kinetic data of the biosorption of Ni^{2+} by QCS at different ionic strengths were analyzed using the pseudo-first order, pseudo-second order, Elovich, fractional power, and intraparticle diffusion models. These models have been used previously to analyze and understand biosorption kinetics of Ni^{2+} ions in single metal systems by different biosorbents [7,11,26,44] but to the best knowledge of the authors, the kinetic process of Ni^{2+} biosorption from aqueous solutions containing NaCl or other background electrolytes has not been mathematically modeled. The parameters of the kinetic models are presented in Table 2.

The highest r^2 values and the lowest SSE, RMSE, and AIC values were obtained using the pseudo-second order, Elovich, and fractional power models. Therefore, these models represent the kinetic profiles of the biosorption of Ni^{2+} by QCS most adequately under the studied conditions.

The equilibrium Ni^{2+} biosorption capacities obtained using the pseudo-second order model (q_{e2}) were similar to those obtained experimentally ($q_{e\ exp}$). Furthermore, as the NaCl ionic strength increased from 0 to 2000 mM, the rate constant of the pseudo-second order model (k_2) decreased from 0.166 to 0.133 g/mmol·h. A similar behavior was observed in the studies of the biosorption of total chromium by QCS at different ionic strengths [20]. A previous study that analyzed the kinetics of the biosorption of Ni^{2+} by QCS at different pH conditions, initial metal concentrations, and temperatures using the pseudo-first order and pseudo-second order models established that the pseudo-second order model described the biosorption kinetics of the heavy metal more satisfactorily than the pseudo-first order model [9]. The present study demonstrates that the pseudo-second order model can also describe the kinetics of the biosorption of Ni^{2+} by QCS at different ionic strengths.

Furthermore, it was observed that higher ionic strengths lead to higher rate constants of the fractional power model (v) and lower initial biosorption rates (A_e), as predicted by the Elovich model, as well as lower constants (k_{FP}) and specific biosorption rates ($k_{FP}·v$) of the fractional power model. The decrease in k_2, A_e, and $k_{FP}·v$ with increasing ionic strengths indicates that the interactions between the QCS biomass and the Ni^{2+} cations decrease with increasing NaCl ionic strengths, which prevents the binding of Ni^{2+} ions to biosorption active sites and decreases the heavy metal biosorption capacity.

Notably, even though NaCl ionic strengths equal to and greater than 200 mM significantly affected the biosorption of Ni^{2+} by QCS, this can be ignored in the present study since the ionic strength of industrial wastewater is lower than 100 mM [45].

Table 2. Effect of the NaCl ionic strength on the parameters of different kinetic models for the biosorption of Ni^{2+} by QCS.

Ionic Strength (mM)	$q_{e\,exp}$ (mmol/g)	Pseudo-First Order Model						Pseudo-Second Order Model					
		k_1 (1/h)	q_{e1} (mmol/g)	r^2	SSE	RMSE	AIC	k_2 (g/mmol·h)	q_{e2} (mmol/g)	r^2	SSE	RMSE	AIC
Control	1.078	0.127 ± 0.041	1.029 ± 0.057	0.9601	0.0710	0.0769	−65.6	0.166 ± 0.054	1.118 ± 0.048	0.9872	0.0234	0.0441	−81.1
0.2	1.030	0.127 ± 0.035	0.998 ± 0.049	0.9771	0.0383	0.0619	−60.0	0.158 ± 0.055	1.079 ± 0.048	0.9892	0.0181	0.0426	−69.0
2	0.910	0.102 ± 0.038	0.863 ± 0.061	0.9422	0.0751	0.0791	−64.8	0.148 ± 0.061	0.953 ± 0.059	0.9775	0.0292	0.0494	−78.0
20	0.798	0.103 ± 0.025	0.776 ± 0.038	0.9772	0.0249	0.0476	−72.7	0.142 ± 0.035	0.860 ± 0.034	0.9916	0.0092	0.0289	−85.6
200	0.510	0.044 ± 0.021	0.512 ± 0.059	0.9229	0.0372	0.0557	−74.6	0.138 ± 0.104	0.565 ± 0.073	0.9521	0.0231	0.0439	−81.3
2000	0.491	0.070 ± 0.037	0.467 ± 0.047	0.9211	0.0339	0.0556	−68.7	0.133 ± 0.084	0.547 ± 0.063	0.9275	0.0183	0.0408	−76.7

	Elovich						Intraparticle Diffusion					
	A_e (mmol/g·h)	B_e (g/mmol)	r^2	SSE	RMSE	AIC	k_{id} (mmol/g·h$^{0.5}$)	c (mmol/g)	r^2	SSE	RMSE	AIC
Control	1.885 ± 1.095	6.577 ± 0.707	0.9964	0.0066	0.0234	−98.85	0.090 ± 0.026	0.259 ± 0.190	0.8227	0.3231	0.1641	−44.37
0.2	1.003 ± 0.531	6.216 ± 0.676	0.9969	0.0052	0.0229	−83.86	0.090 ± 0.023	0.184 ± 0.173	0.8857	0.1910	0.1382	−40.68
2	0.671 ± 0.396	6.728 ± 0.892	0.9931	0.0090	0.0273	−94.56	0.078 ± 0.019	0.188 ± 0.139	0.8679	0.1714	0.1195	−53.24
20	0.274 ± 0.072	6.391 ± 0.470	0.9972	0.0031	0.0167	−99.94	0.072 ± 0.014	0.127 ± 0.099	0.9248	0.0824	0.0866	−57.13
200	0.142 ± 0.088	9.761 ± 1.769	0.9841	0.0076	0.0253	−96.77	0.047 ± 0.006	0.061 ± 0.047	0.9549	0.0217	0.0426	−82.15
2000	0.071 ± 0.040	8.420 ± 1.729	0.9795	0.0088	0.0283	−86.19	0.046 ± 0.005	0.046 ± 0.039	0.9697	0.0130	0.0344	−81.10

	Fractional Power						
	k_{FP} (mmol/g)	v (1/h)	$k_{FP}\cdot v$ (mmol/g·h)	r^2	SSE	RMSE	AIC
Control	0.503 ± 0.040	0.168 ± 0.019	0.0842	0.9963	0.0067	0.0237	−98.58
0.2	0.427 ± 0.038	0.193 ± 0.021	0.0826	0.9975	0.0042	0.0206	−86.41
2	0.360 ± 0.036	0.202 ± 0.024	0.0729	0.9948	0.0068	0.0238	−98.39
20	0.251 ± 0.028	0.256 ± 0.027	0.0641	0.9951	0.0054	0.0222	−92.55
200	0.131 ± 0.020	0.299 ± 0.035	0.0390	0.9926	0.0036	0.0172	−107.5
2000	0.102 ± 0.022	0.342 ± 0.049	0.0350	0.9890	0.0048	0.0208	−94.22

3.2. Influence of Coexisting Ionic Species on the Biosorption of Ni^{2+} by Acorn Shell of Quercus Crassipes Humb. & Bonpl. (QCS)

Generally, there are various types of background electrolytes in industrial wastewater at varying concentrations, which can affect the biosorption of the heavy metal of interest. This study investigated the effect of coexisting ionic species on the biosorption of Ni^{2+} by QCS using different cations and anions at three different concentrations, namely 0.2, 2.0, and 20 mM.

Figure 2 shows the kinetic profiles of the biosorption of Ni^{2+} by QCS at the different concentrations of background electrolytes. There was a small variation in the Ni^{2+} biosorption capacity over the course of the experiments for background electrolyte concentration of 0.2 mM compared to the control experiment (with no background electrolytes) (Figure 2a). However, there were no statistically significant differences in all cases ($p > 0.05$), thus, it is concluded that a background electrolyte concentration of 0.2 mM has no negative effect on the biosorption of Ni^{2+}. Moreover, the global performance indexes for the biosorption of Ni^{2+} at a background electrolyte concentration of 0.2 mM were small and ranged from -7 to -5% (Table 3), which confirms that background electrolytes at a concentration of 0.2 mM have a negligible effect on the biosorption of Ni^{2+}.

Table 3. Effect of background electrolytes at concentrations of 0.2, 2, and 20 mM on the global performance index of the biosorption of Ni^{2+} by QCS.

Background Electrolyte	ξ (%)		
	0.2 mM	2 mM	20 mM
KCl	-6	-8	-21
$NaNO_3$	-6	-12	-20
Na_2SO_4	-5	-10	-27
$CaCl_2$	-6	-20	-54
$MgSO_4$	-7	-23	-77
$MgCl_2$	-7	-24	-42

Increasing the concentration of background electrolytes to 2 mM led to a further decrease in the Ni^{2+} biosorption capacity. However, as shown in Figure 2b, the different background electrolytes influenced the biosorption of Ni^{2+} at varying degrees. The decreases in Ni^{2+} biosorption capacity were more significant in the presence of salts with divalent cations ($CaCl_2$, $MgSO_4$, and $MgCl_2$) and, to a lesser extent, with monovalent cations (KCl, $NaNO_3$, and Na_2SO_4). The statistical analysis revealed that there were significant differences in Ni^{2+} biosorption capacities between the control and the experiments carried out in the presence of compounds with divalent cations ($p < 0.05$). However, there were no statistically significant differences in the maximum Ni^{2+} biosorption capacities between the control and the experiments performed with monovalent cation salts (KCl, $NaNO_3$, and Na_2SO_4) ($p > 0.05$). The global performance indexes corroborated that the compounds containing divalent cations ($CaCl_2$, $MgSO_4$ and $MgCl_2$) affected the biosorption of Ni^{2+} more negatively (from -24% to -20%) compared to the compounds containing monovalent cations (KCl, $NaNO_3$ and Na_2SO_4) (from -12% to -8%).

The background electrolytes at a concentration of 20 mM caused an even greater decrease in the Ni^{2+} biosorption capacity compared to that obtained at 2 mM (Figure 2c). The statistical analysis revealed that the differences between the Ni^{2+} biosorption capacities of the control and the experiments carried out with all the background electrolytes were statistically significant ($p < 0.05$). For all the experiments, the lowest Ni^{2+} biosorption capacities were obtained when $MgSO_4$, $CaCl_2$, and $MgCl_2$ salts were present in the aqueous solutions (Figure 2c). Additionally, the global performance indexes showed that the biosorption of Ni^{2+} was more negatively affected at the highest concentration (20 mM) of background electrolytes and that the negative effect of the background electrolytes followed the order: $MgSO_4 > CaCl_2 > MgCl_2 > Na_2SO_4 > NaNO_3 \approx KCl$.

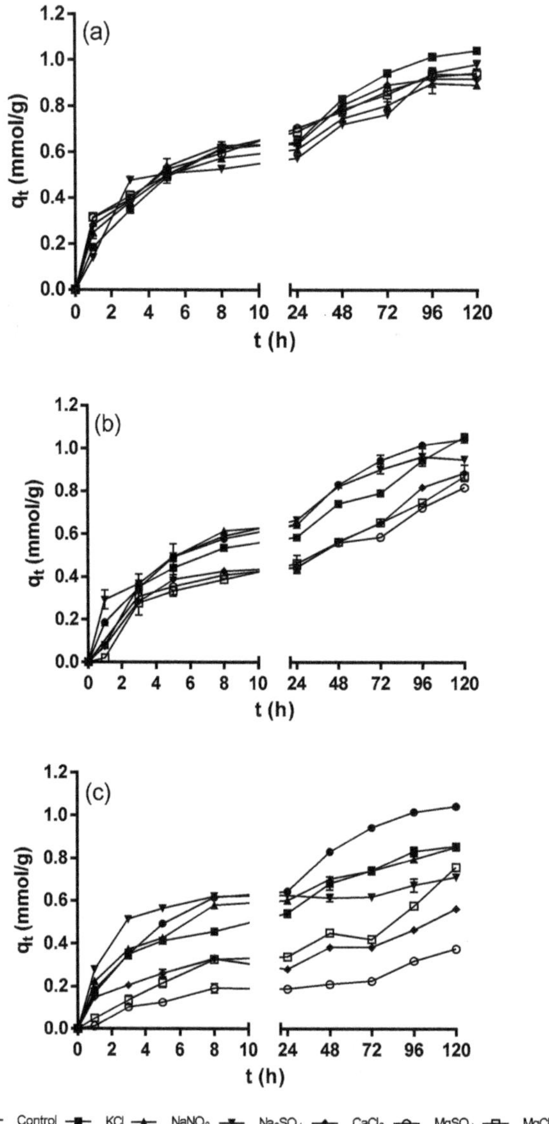

Figure 2. Effect of background electrolytes at a concentration of (**a**) 0.2 mM, (**b**) 2 mM, and (**c**) 20 mM on the biosorption of Ni^{2+} by QCS.

These results clearly show that the Mg^{2+} and Ca^{2+} cations have a more pronounced effect on the biosorption of Ni^{2+} from aqueous solutions by QCS compared to the Na^+ and K^+ cations. This could be attributed to the following reasons: First, the divalent Mg^{2+} and Ca^{2+} cations competed more efficiently with the Ni^{2+} ions for the biosorption active sites present on the QCS surface compared to the monovalent Na^+ and K^+ cations. Second, the divalent Mg^{2+} and Ca^{2+} cations biosorb more easily on the QCS surface compared to the monovalent Na^+ and K^+ cations because of the higher electrostatic attraction. Finally, a Na^+ or K^+ ion biosorbed on the QCS surface occupies only one biosorption active site, whereas a Mg^{2+} or Ca^{2+} ion could occupy two sites, thus resulting in a greater decrease in the biosorption of Ni^{2+} in the presence of $MgCl_2$, $MgSO_4$, and $CaCl_2$ [24,35,46]. These reasons indicate that the biosorption active sites present on the surface of the QCS are not specific to Ni^{2+} ions.

Additionally, the decrease in Ni^{2+} biosorption can be owing to the presence of the Cl^- and SO_4^{2-} ions, which could increase the formation of nickel chloride and nickel sulfate complexes, respectively, thus decreasing the free Ni^{2+} ions in solution [46]. It was observed that the SO_4^{2-} ion had a more negative effect on the biosorption of Ni^{2+} by QCS compared to the Cl^- and NO_3^- ions.

Okoronkwo et al. [46] reported that the Ca^{2+} and Mg^{2+} cations and the SO_4^{2-} anion cause a decrease in the ability of the Mexican sunflower (*Tithonia diversifolia*) stems to bind the Ni^{2+} ions in aqueous solutions. Ca^{2+} and Mg^{2+} cations also decreased the biosorption capacity of *Phellinus igniarius* and *Fomes fomentarius* for Methylene Blue and Rhodamine B dyes [41]. Moreover, Mn^{2+} and Ag^+ ions decreased the Cu^{2+} removal efficiency of granular activated carbon [47]. Likewise, an important decrease in biosorption capacity of *Macrocystis pyrifera* for benzene and toluene was observed, when an artificial seawater solution (Instant Ocean®, Blacksburg, VA, USA) was used, probably because this solution contains different ions [42].

Table 4 presents the equilibrium Ni^{2+} biosorption capacity ($q_{e\ exp}$) for the control experiment (without background electrolytes) and at the different concentrations of background electrolytes. From Table 4, it is evident that the highest biosorption capacity was obtained in the absence of background electrolytes. Moreover, the heavy metal biosorption capacity decreased as the concentration of background electrolytes increased, and the lowest equilibrium Ni^{2+} biosorption capacities were obtained in the presence of $MgSO_4$, $CaCl_2$, and $MgCl_2$ at a concentration of 20 mM. The Elovich, fractional power, and pseudo-second order kinetic models described the kinetic profiles of the biosorption of Ni^{2+} at the three different concentrations of background electrolytes more adequately, as evidenced by the lower SSE, RMSE, and AIC values and the higher r^2 values (Tables 4–6).

The equilibrium Ni^{2+} biosorption capacities predicted by the pseudo-second order model were close to the values obtained experimentally, and they decreased as the concentration of the background electrolytes increased. It is also evident that background electrolytes affected the parameters of all the tested kinetic models, and this occurred to a greater extent at a concentration of 20 mM. Among the observed changes, the decrease in the specific biosorption rate of the fractional power model with increasing concentrations of background electrolytes should be noted.

The goodness of fit between Ni^{2+} biosorption kinetics and pseudo-second order, Elovich, and fractional power models suggests that the biosorption process of Ni^{2+} by QCS from aqueous solutions, containing different types and concentrations of background electrolytes, has a chemical process (chemisorption) as the rate-limiting step of the overall rate of heavy metal biosorption [18,29].

To our knowledge, the effect of ionic strength and background electrolytes on the parameters of kinetic models of the biosorption of Ni^{2+} has not been previously reported.

The results obtained in the present study clearly demonstrate the remarkable capacity of QCS to biosorb Ni^{2+} from aqueous solutions containing various types of impurities with different concentrations.

Additionally, the results of this study show that the impurities in the water greatly affect the performance of the biosorption of heavy metals. Furthermore, this study provides valuable information on the effect of ionic strength, the type, and concentration of background electrolytes on the biosorption of Ni^{2+}, which is of great importance for the application of biosorption technology in the treatment of industrial wastewater contaminated with Ni^{2+}.

Table 4. Parameters of the pseudo-first order and pseudo-second order models for the biosorption of Ni^{2+} by QCS at 0.2, 2, and 20 mM of background electrolytes.

Concentration (mM)	Background Electrolyte	$q_{e\,exp}$ (mmol/g)	Pseudo-First Order Model							Pseudo-Second Order Model					
			k_1 (1/h)	q_{e1} (mmol/g)	r^2	SSE	RMSE	AIC	k_2 (g/mmol·h)	q_{e2} (mmol/g)	r^2	SSE	RMSE	AIC
	Control	1.000	0.142 ± 0.026	0.957 ± 0.047	0.9724	0.0633	0.0629	−94.0	0.170 ± 0.035	1.034 ± 0.041	0.9870	0.0299	0.0433	−107.5
0.2	KCl	0.929	0.180 ± 0.044	0.887 ± 0.058	0.9431	0.0983	0.0784	−86.1	0.280 ± 0.077	0.936 ± 0.047	0.9734	0.0459	0.0536	−99.8
	$NaNO_3$	0.943	0.135 ± 0.042	0.901 ± 0.050	0.9274	0.1452	0.0953	−91.6	0.176 ± 0.079	0.975 ± 0.041	0.9574	0.0852	0.0730	−104.6
	Na_2SO_4	0.937	0.163 ± 0.038	0.902 ± 0.057	0.9493	0.0929	0.0762	−87.1	0.239 ± 0.068	0.957 ± 0.051	0.9724	0.0506	0.0562	−98.0
	$CaCl_2$	0.909	0.193 ± 0.038	0.876 ± 0.046	0.9628	0.0624	0.0624	−94.3	0.308 ± 0.068	0.922 ± 0.036	0.9836	0.0276	0.0415	−108.9
	$MgSO_4$	0.912	0.190 ± 0.049	0.869 ± 0.060	0.9355	0.1054	0.0812	−84.8	0.309 ± 0.094	0.914 ± 0.051	0.9672	0.0537	0.0579	−97.0
	$MgCl_2$	0.909	0.192 ± 0.049	0.870 ± 0.059	0.9371	0.1025	0.0800	−85.3	0.315 ± 0.095	0.914 ± 0.050	0.9679	0.0523	0.0572	−97.4
2	KCl	0.931	0.133 ± 0.040	0.882 ± 0.070	0.9355	0.1397	0.0935	−79.7	0.159 ± 0.065	0.967 ± 0.077	0.9553	0.0968	0.0778	−86.4
	$NaNO_3$	0.865	0.187 ± 0.040	0.829 ± 0.072	0.9529	0.0723	0.0672	−79.1	0.308 ± 0.066	0.875 ± 0.070	0.9771	0.0352	0.0469	−88.6
	Na_2SO_4	0.897	0.181 ± 0.060	0.844 ± 0.075	0.9093	0.1625	0.1008	−77.0	0.272 ± 0.121	0.899 ± 0.074	0.9383	0.1105	0.0831	−84.0
	$CaCl_2$	0.785	0.121 ± 0.035	0.782 ± 0.066	0.9454	0.0732	0.0750	−71.7	0.174 ± 0.071	0.845 ± 0.071	0.9602	0.0533	0.0641	−76.4
	$MgSO_4$	0.709	0.119 ± 0.029	0.765 ± 0.062	0.9644	0.0316	0.0562	−62.2	0.192 ± 0.059	0.812 ± 0.057	0.9795	0.0181	0.0426	−68.9
	$MgCl_2$	0.757	0.108 ± 0.033	0.755 ± 0.065	0.9465	0.0696	0.0732	−72.4	0.153 ± 0.065	0.824 ± 0.073	0.9602	0.0517	0.0631	−76.9
20	KCl	0.810	0.152 ± 0.052	0.737 ± 0.063	0.8968	0.1490	0.0910	−90.5	0.230 ± 0.088	0.811 ± 0.058	0.9499	0.0724	0.0634	−104.9
	$NaNO_3$	0.795	0.204 ± 0.054	0.737 ± 0.048	0.9309	0.0921	0.0715	−100.1	0.350 ± 0.104	0.793 ± 0.041	0.9666	0.0445	0.0497	−114.6
	Na_2SO_4	0.667	0.524 ± 0.090	0.642 ± 0.020	0.9779	0.0190	0.0325	−131.7	1.325 ± 0.355	0.671 ± 0.023	0.9766	0.0201	0.0334	−130.5
	$CaCl_2$	0.470	0.212 ± 0.072	0.427 ± 0.040	0.9031	0.0383	0.0505	−95.8	0.698 ± 0.335	0.451 ± 0.040	0.9287	0.0282	0.0433	−101.0
	$MgSO_4$	0.307	0.211 ± 0.054	0.211 ± 0.016	0.9576	0.0040	0.0175	−115.4	1.158 ± 0.572	0.233 ± 0.023	0.9486	0.0048	0.0192	−112.5
	$MgCl_2$	0.584	0.047 ± 0.024	0.627 ± 0.098	0.8653	0.1335	0.0913	−80.6	0.089 ± 0.063	0.715 ± 0.117	0.9117	0.0875	0.0739	−88.1

Table 5. Parameters of the Elovich and intraparticle diffusion models for the biosorption of Ni^{2+} by QCS at 0.2, 2, and 20 mM of background electrolytes.

Concentration (mM)	Background Electrolyte	Elovich						Intraparticle Diffusion					
		A_e (mmol/g·h)	B_e (g/mmol)	r^2	SSE	RMSE	AIC	k_{id} (mmol/g·h$^{0.5}$)	c (mmol/g)	r^2	SSE	RMSE	AIC
	Control	0.418 ± 0.110	5.369 ± 0.418	0.9916	0.0193	0.0348	−115.3	0.087 ± 0.014	0.183 ± 0.089	0.9126	0.2006	0.1120	−73.22
0.2	KCl	0.918 ± 0.196	6.937 ± 0.357	0.9960	0.0069	0.0208	−133.9	0.074 ± 0.014	0.236 ± 0.091	0.8803	0.2067	0.1137	−72.69
	$NaNO_3$	0.856 ± 0.304	7.391 ± 0.632	0.9891	0.0168	0.0324	−117.9	0.069 ± 0.015	0.222 ± 0.091	0.8641	0.2088	0.1142	−72.51
	Na_2SO_4	0.738 ± 0.248	6.552 ± 0.560	0.9890	0.0201	0.0354	−114.7	0.077 ± 0.015	0.224 ± 0.092	0.8840	0.2124	0.1152	−72.20
	$CaCl_2$	1.051 ± 0.392	7.210 ± 0.625	0.9888	0.0188	0.0343	−115.8	0.072 ± 0.016	0.245 ± 0.102	0.8437	0.2623	0.1280	−68.40
	$MgSO_4$	1.078 ± 0.331	7.306 ± 0.518	0.9925	0.0123	0.0278	−123.4	0.072 ± 0.015	0.243 ± 0.092	0.8690	0.2141	0.1157	−72.06
	$MgCl_2$	1.128 ± 0.355	7.362 ± 0.529	0.9922	0.0126	0.0281	−123.0	0.071 ± 0.015	0.247 ± 0.094	0.8643	0.2211	0.1175	−71.48
2	KCl	0.284 ± 0.139	5.350 ± 0.845	0.9689	0.0674	0.0649	−92.85	0.085 ± 0.013	0.135 ± 0.084	0.9187	0.1760	0.1049	−75.58
	$NaNO_3$	0.409 ± 0.137	5.765 ± 0.568	0.9863	0.0275	0.0414	−109.0	0.082 ± 0.011	0.172 ± 0.071	0.9372	0.1257	0.0886	−81.65
	Na_2SO_4	0.600 ± 0.407	6.655 ± 1.190	0.9548	0.0810	0.0712	−89.55	0.075 ± 0.016	0.197 ± 0.098	0.8653	0.2412	0.1228	−69.21
	$CaCl_2$	0.257 ± 0.135	6.352 ± 1.061	0.9715	0.0382	0.0542	−81.42	0.070 ± 0.012	0.125 ± 0.077	0.9223	0.1041	0.0895	−66.38
	$MgSO_4$	0.342 ± 0.117	7.261 ± 0.740	0.9920	0.0071	0.0266	−80.22	0.062 ± 0.014	0.151 ± 0.084	0.9126	0.0774	0.0880	−51.52
	$MgCl_2$	0.191 ± 0.098	6.178 ± 1.074	0.9709	0.0379	0.0540	−81.55	0.069 ± 0.011	0.100 ± 0.071	0.9329	0.0872	0.0819	−69.02
20	KCl	0.399 ± 0.128	6.997 ± 0.418	0.9865	0.0196	0.0330	−131.1	0.069 ± 0.010	0.167 ± 0.064	0.9142	0.1238	0.0830	−94.19
	$NaNO_3$	0.815 ± 0.334	8.064 ± 0.639	0.9839	0.0214	0.0345	−129.3	0.064 ± 0.014	0.217 ± 0.083	0.8440	0.2077	0.1074	−83.84
	Na_2SO_4	0.425 ± 0.587	15.69 ± 0.790	0.9219	0.0675	0.0612	−106.3	0.041 ± 0.019	0.323 ± 0.116	0.5324	0.4025	0.1495	−70.61
	$CaCl_2$	0.576 ± 0.479	14.73 ± 1.665	0.9505	0.0195	0.0361	−107.2	0.037 ± 0.009	0.120 ± 0.054	0.8315	0.0665	0.0666	−86.41
	$MgSO_4$	0.163 ± 0.170	24.01 ± 7.875	0.9113	0.0082	0.0253	−104.3	0.024 ± 0.009	0.050 ± 0.039	0.7443	0.0239	0.0429	−88.45
	$MgCl_2$	0.074 ± 0.041	6.100 ± 1.519	0.9453	0.0542	0.0582	−96.78	0.062 ± 0.009	0.040 ± 0.042	0.9486	0.0509	0.0564	−97.91

Table 6. Parameters of the fractional power model for the biosorption of Ni^{2+} by QCS at 0.2, 2, and 20 mM of background electrolytes.

Concentration (mM)	Background Electrolyte	k_{FP} (mmol/g)	v (1/h)	$k_{FP}v$ (mmol/g·h)	r^2	SSE	RMSE	AIC
				Fractional Power				
	Control	0.289 ± 0.043	0.275 ± 0.036	0.0793	0.9768	0.0534	0.0578	−97.02
0.2	KCl	0.340 ± 0.025	0.221 ± 0.018	0.0753	0.9906	0.0162	0.0318	−118.6
	NaNO$_3$	0.320 ± 0.036	0.220 ± 0.028	0.0705	0.9786	0.0329	0.0453	−105.8
	Na$_2$SO$_4$	0.327 ± 0.035	0.233 ± 0.027	0.0761	0.9825	0.0321	0.0448	−106.2
	CaCl$_2$	0.352 ± 0.041	0.210 ± 0.029	0.0741	0.9755	0.0411	0.0507	−101.8
	MgSO$_4$	0.347 ± 0.027	0.213 ± 0.020	0.0738	0.9887	0.0185	0.0340	−116.1
	MgCl$_2$	0.351 ± 0.028	0.210 ± 0.020	0.0738	0.9879	0.0197	0.0351	−115.0
2	KCl	0.232 ± 0.053	0.308 ± 0.055	0.0714	0.9579	0.0911	0.0755	−87.43
	NaNO$_3$	0.269 ± 0.028	0.278 ± 0.026	0.0747	0.9881	0.0237	0.0385	−111.6
	Na$_2$SO$_4$	0.295 ± 0.061	0.246 ± 0.051	0.0724	0.9442	0.0999	0.0790	−85.77
	CaCl$_2$	0.209 ± 0.047	0.294 ± 0.053	0.0614	0.9659	0.0457	0.0593	−78.71
	MgSO$_4$	0.236 ± 0.021	0.254 ± 0.022	0.0598	0.9942	0.0051	0.0226	−84.11
	MgCl$_2$	0.179 ± 0.045	0.319 ± 0.059	0.0572	0.9646	0.0461	0.0595	−78.6
20	KCl	0.245 ± 0.025	0.264 ± 0.026	0.0645	0.9845	0.0224	0.0351	−128.4
	NaNO$_3$	0.301 ± 0.035	0.216 ± 0.030	0.0652	0.9719	0.0374	0.0456	−118.1
	Na$_2$SO$_4$	0.423 ± 0.059	0.107 ± 0.038	0.0451	0.9078	0.0794	0.0664	−103.1
	CaCl$_2$	0.175 ± 0.031	0.213 ± 0.045	0.0372	0.9472	0.0208	0.0373	−106.1
	MgSO$_4$	0.085 ± 0.027	0.239 ± 0.093	0.0204	0.8746	0.0117	0.0300	−99.14
	MgCl$_2$	0.101 ± 0.029	0.404 ± 0.069	0.0406	0.9564	0.0432	0.0520	−100.9

4. Conclusions

The results obtained in this study demonstrated that QCS is a useful biosorbent for the bioremediation of aqueous solutions contaminated with Ni^{2+} and containing inorganic impurities. The effect of background electrolytes varied depending on the type and concentration of the electrolyte. The biosorption of Ni^{2+} was not affected by a NaCl ionic strength of 0.2 mM but was affected by higher ionic strengths. In addition, the background electrolytes (KCl, $NaNO_3$, Na_2SO_4, $CaCl_2$, $MgSO_4$, and $MgCl_2$) had no effect on the biosorption of the heavy metal at a concentration of 0.2 mM. However, at concentrations of 2 and 20 mM, the divalent Ca^{2+} and Mg^{2+} cations affected the biosorption of Ni^{2+} significantly, whereas the monovalent Na^+ and K^+ cations affected the biosorption of Ni^{2+} only slightly. The kinetic experimental data were well represented using the Elovich, fractional power, and pseudo-second order models. In order to determine the full potential of QCS as a commercial biosorbent, Ni^{2+} biosorption from industrial wastewater will be evaluated in a future research study.

Author Contributions: Conceptualization, E.C.-U.; methodology, E.A.-G., G.M.C.-C., E.C.-U.; software, E.A.-G.; validation, E.A.-G.; formal analysis, E.A.-G., G.M.C.-C., E.C.-U.; investigation, E.A.-G.; resources, E.C.-U.; writing—original draft preparation, E.A.-G., G.M.C.-C., E.C.-U.; writing—review and editing, E.A.-G., G.M.C.-C., E.C.-U.; visualization, E.A.-G.; supervision, E.C.-U.; project administration, E.C.-U.; funding acquisition, E.C.-U. All authors have read and agreed to the published version of the manuscript.

Funding: This research was funded by the Instituto Politécnico Nacional, Secretaría de Investigación y Posgrado, project number: SIP20201439.

Acknowledgments: E.C.-U. and G.M.C.-C. hold grants from EDI-IPN, COFAA-IPN, and SNI-CONACYT. E.A.-G. holds a grant from SNI-CONACYT.

Conflicts of Interest: The authors declare no conflict of interest.

References

1. Barquilha, C.E.; Cossich, E.S.; Tavares, C.R.; Da Silva, E.A. Biosorption of nickel and copper ions from synthetic solution and electroplating effluent using fixed bed column of immobilized brown algae. *J. Water Process. Eng.* **2019**, *32*, 100904. [CrossRef]

2. Lopez-Nuñez, P.V.; Aranda-Garcia, E.; Cristiani-Urbina, M.D.C.; Morales-Barrera, L.; Cristiani-Urbina, E. Removal of hexavalent and total chromium from aqueous solutions by Plum (*P. domestica* L.) tree bark. *Environ. Eng. Manag. J.* **2014**, *13*, 1927–1938. [CrossRef]

3. Sharma, R.; Singh, B. Removal of Ni (II) ions from aqueous solutions using modified rice straw in a fixed bed column. *Bioresour. Technol.* **2013**, *146*, 519–524. [CrossRef] [PubMed]

4. Altino, H.O.N.; Costa, B.E.; Da Cunha, R.N. Biosorption optimization of Ni(II) ions on Macauba (*Acrocomia aculeata*) oil extraction residue using fixed-bed column. *J. Environ. Chem. Eng.* **2017**, *5*, 4895–4905. [CrossRef]

5. Orhan, Y.; Hrenovič, J.; Büyükgüngör, H. Biosorption of heavy metals from wastewater by biosolids. *Eng. Life Sci.* **2006**, *6*, 399–402. [CrossRef]

6. Coman, V.; Robotin, B.; Ilea, P. Nickel recovery/removal from industrial wastes: A review. *Resour. Conserv. Recycl.* **2013**, *73*, 229–238. [CrossRef]

7. Flores-Garnica, J.G.; Morales-Barrera, L.; Pineda-Camacho, G.; Cristiani-Urbina, E. Biosorption of Ni(II) from aqueous solutions by *Litchi chinensis* seeds. *Bioresour. Technol.* **2013**, *136*, 635–643. [CrossRef]

8. Bobadilla, M.C.; Lostado-Lorza, R.; Gómez, F.S.; Escribano-Garcia, R. Adsorptive of Nickel in wastewater by olive stone waste: Optimization through multi-response surface methodology using desirability functions. *Water* **2020**, *12*, 1320. [CrossRef]

9. Aranda-García, E.; Cristiani-Urbina, E. Kinetic, equilibrium, and thermodynamic analyses of Ni(II) biosorption from aqueous solution by acorn shell of *Quercus crassipes*. *Water Air Soil Pollut.* **2018**, *229*, 119. [CrossRef]

10. Pandey, P.K.; Choubey, S.; Verma, Y.; Pandey, M.; Kamal, S.S.K.; Chandrashekhar, K. Biosorptive removal of Ni(II) from wastewater and industral effluent. *Int. J. Environ. Res. Public Health* **2007**, *4*, 332–339. [CrossRef]

11. Suazo-Madrid, A.; Morales-Barrera, L.; Aranda-García, E.; Cristiani-Urbina, E. Ni(II) biosorption by *Rhodotorula glutinis*. *J. Ind. Microbiol. Biotechnol.* **2011**, *38*, 51–64. [CrossRef] [PubMed]

12. World Health Organization. *Guidelines for Drinking-Water Quality*, 4th ed.; WHO: Geneva, Switzerland, 2017.

13. Villen-Guzman, M.; Gutierrez-Pinilla, D.; Gomez-Lahoz, C.; Vereda-Alonso, C.; Rodriguez-Maroto, J.; Arhoun, B. Optimization of Ni (II) biosorption from aqueous solution on modified lemon peel. *Environ. Res.* **2019**, *179*, 108849. [CrossRef]

14. Beni, A.A.; Esmaeili, A. Biosorption, an efficient method for removing heavy metals from industrial effluents: A review. *Environ. Technol. Innov.* **2020**, *17*, 100503. [CrossRef]

15. Ali, I.H.; Al Mesfer, M.K.; Khan, M.I.; Danish, M.; Alghamdi, M.M. Exploring adsorption process of Lead (II) and Chromium (VI) ions from aqueous solutions on acid activated carbon prepared from *Juniperus procera* leaves. *Processes* **2019**, *7*, 217. [CrossRef]

16. Fomina, M.; Gadd, G.M. Biosorption: Current perspectives on concept, definition and application. *Bioresour. Technol.* **2014**, *160*, 3–14. [CrossRef]

17. Kumar, N.S.; Asif, M.; Poulose, A.M.; Suguna, M.; Al-Hazza, M.I. Equilibrium and kinetic studies of biosorptive removal of 2,4,6-trichlorophenol from aqueous solutions using untreated agro-waste pine cone biomass. *Processes* **2019**, *7*, 757. [CrossRef]

18. Aranda-García, E.; Cristiani-Urbina, E. Effect of pH on hexavalent and total chromium removal from aqueous solutions by avocado shell using batch and continuous systems. *Environ. Sci. Pollut. Res.* **2017**, *26*, 3157–3173. [CrossRef]

19. Netzahuatl-Muñoz, A.R.; Cristiani-Urbina, M.D.C.; Cristiani-Urbina, E. Chromium biosorption from Cr(VI) aqueous solutions by *Cupressus lusitanica* bark: Kinetics, equilibrium and thermodynamic studies. *PLoS ONE* **2015**, *10*, e0137086. [CrossRef]

20. Aranda-García, E.; Morales-Barrera, L.; Pineda-Camacho, G.; Cristiani-Urbina, E. Effect of pH, ionic strength, and background electrolytes on Cr(VI) and total chromium removal by acorn shell of *Quercus crassipes* Humb. & Bonpl. *Environ. Monit. Assess.* **2014**, *186*, 6207–6221. [CrossRef]

21. Aranda-García, E.; Netzahuatl-Muñoz, A.R.; Cristiani-Urbina, M.; Morales-Barrera, L.; Pineda-Camacho, G.; Cristiani-Urbina, E. Bioreduction of Cr(VI) and chromium biosorption by acorn shell of *Quercus crassipes* Humb. & Bonpl. *J. Biotechnol.* **2010**, *150*, 228. [CrossRef]

22. Aranda-García, E.; Cristiani-Urbina, E. Hexavalent chromium removal and total chromium biosorption from aqueous solution by *Quercus crassipes* acorn shell in a continuous up-flow fixed-bed column: Influencing parameters, kinetics, and mechanism. *PLoS ONE* **2020**, *15*, e0227953. [CrossRef]

23. Mack, C.; Wilhelmi, B.; Duncan, J.; Burgess, J. Biosorption of precious metals. *Biotechnol. Adv.* **2007**, *25*, 264–271. [CrossRef]

24. Hu, X.-J.; Liu, Y.; Zeng, G.-M.; You, S.-H.; Wang, H.; Hu, X.; Guo, Y.-M.; Tan, X.; Guo, F.-Y. Effects of background electrolytes and ionic strength on enrichment of Cd(II) ions with magnetic graphene oxide–supported sulfanilic acid. *J. Colloid Interface Sci.* **2014**, *435*, 138–144. [CrossRef] [PubMed]

25. Kidgell, J.T.; De Nys, R.; Hu, Y.; Paul, N.A.; Roberts, D.A. Bioremediation of a complex industrial effluent by biosorbents derived from freshwater macroalgae. *PLoS ONE* **2014**, *9*, e94706. [CrossRef]

26. Hernández-Estévez, A.; Cristiani-Urbina, E. Nickel(II) biosorption from aqueous solutions by shrimp head biomass. *Environ. Monit. Assess.* **2014**, *186*, 7987–7998. [CrossRef]

27. Mitchell, A.; Mellon, M. Colorimetric determination of nickel with dimethylglyoxime. *Ind. Eng. Chem. Anal. Ed.* **1945**, *17*, 380–382. [CrossRef]

28. Albadarin, A.B.; Al-Muhtaseb, A.H.; Al-Laqtah, N.A.; Walker, G.; Allen, S.; Ahmad, M.N. Biosorption of toxic chromium from aqueous phase by lignin: Mechanism, effect of other metal ions and salts. *Chem. Eng. J.* **2011**, *169*, 20–30. [CrossRef]

29. Febrianto, J.; Kosasih, A.N.; Sunarso, J.; Ju, Y.-H.; Indraswati, N.; Ismadji, S. Equilibrium and kinetic studies in adsorption of heavy metals using biosorbent: A summary of recent studies. *J. Hazard. Mater.* **2009**, *162*, 616–645. [CrossRef]

30. Plazinski, W.; Rudzinski, W.; Plazinska, A. Theoretical models of sorption kinetics including a surface reaction mechanism: A review. *Adv. Colloid Interface Sci.* **2009**, *152*, 2–13. [CrossRef]

31. Basha, S.; Murthy, Z. Kinetic and equilibrium models for biosorption of Cr(VI) on chemically modified seaweed, *Cystoseira indica*. *Process Biochem.* **2007**, *42*, 1521–1529. [CrossRef]

32. Subbaiah, M.V.; Vijaya, Y.; Kumar, N.S.; Reddy, A.S.; Abburi, K. Biosorption of nickel from aqueous solutions by *Acacia leucocephala* bark: Kinetics and equilibrium studies. *Colloids Surfaces B Biointerfaces* **2009**, *74*, 260–265. [CrossRef] [PubMed]

33. Schiewer, S.; Volesky, B. Ionic strength and electrostatic effects in biosorption of divalent metal ions and protons. *Environ. Sci. Technol.* **1997**, *31*, 2478–2485. [CrossRef]

34. Barnie, S.; Zhang, J.; Wang, H.; Yin, H.; Chen, H. The influence of pH, co-existing ions, ionic strength, and temperature on the adsorption and reduction of hexavalent chromium by undissolved humic acid. *Chemosphere* **2018**, *212*, 209–218. [CrossRef] [PubMed]

35. Villaescusa, I.; Fiol, N.; Martínez, M.; Miralles, N.; Poch, J.; Serarols, J. Removal of copper and nickel ions from aqueous solutions by grape stalks wastes. *Water Res.* **2004**, *38*, 992–1002. [CrossRef]

36. Mogollón, L.; Rodriguez, R.; Larrota, W.; Ramírez, N.; Torres, R. Biosorption of nickel using filamentous fungi. *Appl. Biochem. Biotechnol.* **1998**, *70–72*, 593–601. [CrossRef]

37. Thevannan, A.; Mungroo, R.; Niu, C.H. Biosorption of nickel with barley straw. *Bioresour. Technol.* **2010**, *101*, 1776–1780. [CrossRef]

38. Djemmoe, L.G.; Njanja, T.E.; Deussi, M.C.N.; Tonle, K.I. Assessment of copper(II) biosorption from aqueous solution by agricultural and industrial residues. *Comptes Rendus Chim.* **2016**, *19*, 841–849. [CrossRef]

39. Zhou, W.; Wang, J.; Shen, B.; Hou, W.; Zhang, Y. Biosorption of copper(II) and cadmium(II) by a novel exopolysaccharide secreted from deep-sea mesophilic bacterium. *Colloids Surf. B Biointerfaces* **2009**, *72*, 295–302. [CrossRef]

40. Verma, A.; Kumar, S.; Kumar, S. Biosorption of lead ions from the aqueous solution by *Sargassum filipendula*: Equilibrium and kinetic studies. *J. Environ. Chem. Eng.* **2016**, *4*, 4587–4599. [CrossRef]

41. Maurya, N.S.; Mittal, A.K.; Cornel, P.; Rother, E. Biosorption of dyes using dead macro fungi: Effect of dye structure, ionic strength and pH. *Bioresour. Technol.* **2006**, *97*, 512–521. [CrossRef]

42. Flores-Chaparro, C.E.; Ruiz, L.F.C.; De La Torre, M.C.A.; Huerta-Diaz, M.A.; Rangel-Mendez, J.R. Biosorption removal of benzene and toluene by three dried macroalgae at different ionic strength and temperatures: Algae biochemical composition and kinetics. *J. Environ. Manag.* **2017**, *193*, 126–135. [CrossRef] [PubMed]

43. Agarwal, A.; Upadhyay, U.; Sreedhar, I.; Singh, S.A.; Patel, C.M. A review on valorization of biomass in heavy metal removal from wastewater. *J. Water Process. Eng.* **2020**, *38*, 101602. [CrossRef]

44. Vázquez-Palma, D.E.; Netzahuatl-Muñoz, A.R.; Pineda-Camacho, G.; Cristiani-Urbina, E. Biosorptive removal of nickel(II) ions from aqueous solutions by Hass avocado (*Persea americana* Mill. var. Hass) shell as an effective and low-cost biosorbent. *Fresenius Environ. Bull.* **2017**, *26*, 3501–3513.

45. Park, D.; Yun, Y.S.; Jo, J.H.; Park, J.M. Effects of ionic strength, background electrolytes, heavy metals, and redox-active species on the reduction of hexavalent chromium by *Ecklonia* biomass. *J. Microbiol. Biotechnol.* **2005**, *15*, 780–786.

46. Okoronkwo, A.E.; Aiyesanmi, A.F.; Olasehinde, E.F. Biosorption of nickel from aqueous solution by *Tithonia diversifolia*. *Desalin. Water Treat.* **2009**, *12*, 352–359. [CrossRef]

47. Almohammadi, S.; Mirzaei, M. Removal of copper (II) from aqueous solutions by adsorption onto granular activated carbon in the presence of competitor ions. *Adv. Environ. Technol.* **2016**, *2*, 85–94.

 *processes*

Review

Biosorption: A Review of the Latest Advances

Enrique Torres

Laboratorio de Microbiología, Facultad de Ciencias, Universidade da Coruña, Campus de A Zapateira,
15071 A Coruña, Spain; enrique.torres@udc.es; Tel.: +34-981-167-000; Fax: +34-981-167-065

Received: 14 September 2020; Accepted: 27 November 2020; Published: 1 December 2020

Abstract: Biosorption is a variant of sorption techniques in which the sorbent is a material of biological origin. This technique is considered to be low cost and environmentally friendly, and it can be used to remove pollutants from aqueous solutions. The objective of this review is to report on the most significant recent works and most recent advances that have occurred in the last couple of years (2019–2020) in the field of biosorption. Biosorption of metals and organic compounds (dyes, antibiotics and other emerging contaminants) is considered in this review. In addition, the use and possibilities of different forms of biomass (live or dead, modified or immobilized) are also considered.

Keywords: biosorption; biomass; bioaccumulation; biocomposite; pollutants; metals; emerging organic contaminants; dyes

1. Introduction

Biosorption is a variant of the sorption techniques in which the sorbent is a material of biological origin. Today, biosorption is considered a simple, economical and environmentally friendly process that is used as an attractive alternative for removing pollutants. Within this context, biosorption is a general term that describes the removal of pollutants by their binding to a material of biological origin (biomass). There have been numerous biosorption studies in the last decade, and advances in this field have reinforced the interest in this technique to solve environmental pollution problems. The existing information on biosorption is abundant due to the large number of works that are published to test the validity of certain materials as biosorbents or to develop more complex hybrid materials that can be more efficient for this purpose. This review aims to evaluate the latest contributions (in the last couple of years: 2019–August 2020) in the field of biosorption. Although biosorption is a mechanism that also acts in soil decontamination, this review will focus on biosorption processes from aqueous solutions.

2. Biosorption: Generalities

The main element of a biosorption process is biomass. The term biomass is a very broad term that includes intact living cells and derived compounds of biological origin with different degrees of transformation (waste, charcoal, etc.). Taking this into account, the use of biological materials as sorbents has an important alternative: this biomass can be alive or dead. In the case of dead biomass, the pollutants passively (metabolism-independent) bind to this type of biomass through ionic, chemical or physical mechanisms (biosorption); however, with living biomass, the process is more complex because the metabolic activity of this biomass is added to the passive mechanisms. This metabolic activity allows the active transport of pollutants through the membrane into the cell interior. In this way, pollutants can accumulate inside the cell (bioaccumulation). Furthermore, since the enzymatic activity is preserved in living biomass, there is also the possibility that different enzymatic activities may alter the state of the pollutant (biodegradation and biotransformation).

Considering the previous information, the use of living biomass as a biosorbent would have more possibilities to remove a greater amount of pollutants, which constitutes an important advantage when using this type of biomass. However, other advantages and disadvantages must be considered in the use of one type or another of biomass. Most applications focus on the use of dead biomass because toxicity related problems are avoided, no maintenance is required, this biomass can be stored for long periods without loss of effectiveness, regeneration is more feasible and it is possible to work on a greater range of environmental variables. In addition, this biomass can be cut and ground to obtain a suitable particle size. However, despite all these advantages of dead biomass, the use of living biomass can have an important advantage, since, as indicated above, the cells are metabolically active, so the pollutants can be incorporated into the cell interior increasing the efficiency of the process because bioaccumulation contributes to the initial biosorption process [1,2]. In this case, there would be a first step, independent of metabolism, in which the pollutant would bind to the cell surface (biosorption in the strict sense), and a second step, dependent on metabolism, in which the pollutant is transported through the cell membrane to the cell interior. At this step, it must be taken into account that some pollutants could also pass through the membrane by passive diffusion. In many cases, the term biosorption is used in a general way to include both steps when using living systems, although both steps are different.

Continuing with the advantages and disadvantages of living biomass, an additional advantage that active systems have is that there is also the possibility of biotransformation or biodegradation, increasing, in some cases, the ability to eliminate a higher amount of pollutant [3–7]. However, there are also disadvantages that can be attributed to the use of living biomass. Thus, it must be considered that to use living biomass it is necessary to have culture systems, nutrient supply and some method for cell harvesting, which makes the process more expensive. However, dead biomass can also have additional costs, since, in some cases, this type of biomass is chemically modified, carbonized or ground to make it more effective, which does not apply to living biomass. Additionally, living biomass is easier to separate from a reaction system.

In any case, an important aspect to consider in order to achieve the advantages of living biomass can be effective is to look for those organisms that show greater resistance to the toxic effects of the pollutant. Hence, this is a first step to optimize a biosorption/bioaccumulation process using living biomass. Strains more resistant to the target pollutant can have a greater removal capacity; a recent example is the use of a strain of *Pseudomonas* sp. with resistances to multiple heavy metals for cadmium removal; this resistant strain used as living biomass was more effective than the dead one [8]. In this sense, there is currently a growing interest in the use of microorganisms as base material to develop biosorbents due to their good sorption properties and resistance to the toxic effect of pollutants. Various species of fungi, bacteria, yeast and microalgae have been tested against many types of pollutants with very promising results [4,8–13].

In addition to biomass from microorganisms, as indicated above, a large number of materials have been evaluated as biosorbents to eliminate different pollutants, among the most recent are: agro-industrial waste materials [14], sludge [15,16], polysaccharides [17], plant-derived materials [18–20] and biopolymers [21]; although it is necessary to indicate that throughout the years of development of biosorption, a large number of biological materials of very diverse origin have been evaluated as possible biosorbents [22,23]. Many of these materials are considered as waste: for this reason, the use of these materials as biosorbents has a double advantage, on the one hand, a waste is used for an application, and therefore, its waste is reduced; on the other hand, this material is used to eliminate pollution with a possible low cost.

These biomaterials can be applied directly or immobilized on different supports. Recently, there has been an increase in the number of studies applying biomaterials packed in fixed-bed columns [24,25]. This technique offers advantages for practical applications of large-scale biosorption processes because it allows continuous work. Thus, to operate with these columns, the biomass must be immobilized, which is necessary when using biomass from microorganisms. A common alternative (which can even be used with living cells) is to immobilize the biosorbent in a calcium alginate matrix [6,26,27]. Immobilization allows the biomass to be retained in a reactor, reduces separation costs and increases the mechanical resistance of the biomass. In this context, although biosorption/bioaccumulation techniques are usually simple, more complex systems have now been developed, forming biocomposite materials with new characteristics. These materials with polymeric structures allow to protect and maintain the viability of living systems, which makes it possible to take advantage of the highest efficiency of these systems. An example of this was the use of biomass from *Lysinibacillus sphaericus* CBAM5 immobilized in polycaprolactone microfibrous mats and alginate microcapsules to capture gold from synthetic water samples [28]. Although the application of nanoparticles in the field of biosorption has been developing for a long time, more recently the application of magnetic nanoparticles as a support for the immobilization of microorganisms has also been assessed, as evidenced in the review by Giese et al. (2020) [29].

Another way to use biomass is by its chemical modification to increase the sorption capacity. These modifications alter the functional groups of the biomass and its surface topography favoring the binding of pollutants. Some examples have recently been published in which different modification methods are used, such as esterification, graft polymerization, coating, treatments with acids, alkalis, methanol, cationic surfactants, formaldehyde or triethyl phosphate and nitromethane [30–33]. Although these modifications apply to dead biomass, living biomass can also be modified, but in a very different way, by genetic modification, which allows the introduction of genes into the desired biomass that increase resistance to the toxic effect of certain pollutants, or that increase the uptake of the pollutant (several examples with metals have recently been published) [34–36].

Although biosorption alone is an effective technique, its flexibility allows coupling with other techniques. For example, biosorption allowed to replace expensive materials used in anodic oxidation processes with plant material [37]; in this way, hybrid materials are created increasing the efficiency of biosorption [38].

Since biosorption requires an interaction between the biomass and the pollutant (usually multiple interactions coexist), those factors that influence this interaction will influence the efficiency of the process. Interactions such as ion exchange, complexation/coordination, electrostatic interactions, chemisorption, physisorption, microprecipitation and reduction can be established in a biosorption process. Taking this into account, factors such as pH, temperature (thermodynamic studies), contact time, shaking speed, initial concentration of the pollutant or amount of biosorbent are well known and are evaluated to optimize the biosorption process [23]. However, there are other less studied factors that have an impact on the process—for example, the type and amount of functional groups in the biomass such as carboxyl, amino, phosphoryl or sulfonate and that influence the biosorption of some pollutants (mainly metals) [9,31,39], ionic strength [40], presence of dissolved organic matter that alters the absorption of metals [41] or the competition with other pollutants [40,42,43]. When using a living system, it is necessary to consider other factors such as the response to the possible toxicity of the pollutant [44,45].

It is evident that the nature of the biosorbent determines many of its physico-chemical properties, such as the type and quantity of functional groups, but there are other aspects that are also especially relevant in the process, such as surface area (increasing the surface area increases the contact of the sorbate with the sorbent), porosity (mesoporosity increases the biosorption capacity, while microporosity decreases it) or cell structures (different structures have different physico-chemical characteristics) [39,46,47]. The characterization of the material in terms of its morphology and composition are common in biosorption studies because they allow a more detailed description and provide information on the sorption mechanisms. Fourier transform infrared spectroscopy (FTIR), scanning electron microscopy (SEM) and energy dispersive x-ray spectroscopy (EDS) are methods commonly used for this purpose. Thermogravimetric test, elemental analysis, zeta potential and point of zero charge (pHzpc) measurement are alternative parameters that also provide information on the physico-chemical characteristics of the biosorbent. Recently, an electrokinetic method termed as dielectrophoresis (DEP) was applied for the characterization of biosorption [48].

Kinetic and isotherm models are used for the mathematical modeling of the biosorption processes. Traditionally, the most widely used kinetic models are pseudo-first order [49], pseudo-second order [50] and intraparticle [51]. From these models, different parameters are obtained that allow characterizing the speed of the process, time to reach equilibrium or even determine which stages of the process may be limiting. Currently there are other kinetic models that are being used and that are useful to characterize biosorption processes such as the kinetics Brouers-Sotolongo family model [38,52]. Regarding the isotherm models, the Langmuir [53] and Freundlich [54] models are still the most widely used.

The search and evaluation of new biosorbents is a current challenge in this field. These new biosorbents must be more efficient, economical and have good reusability through various application cycles. Although this last property is desired for many sorbents, in the case of biosorbents (used as dead biomass) it can be considered non-essential because, precisely, a property of these materials must be their low cost. Instead, it is important that these materials prove their worth under real application conditions.

Everything said above shows the great interest in applying biosorption-related techniques to solve pollution problems. However, the interest of biosorption goes beyond its usefulness as a pollutant remover: biosorption techniques even allow the recovery of some useful substances. Since many of the mechanisms involved in biosorption are reversible, there is the possibility of recovering the biosorbed materials, and at the same time regenerating the biosorbent. An example of this application is the rare-earths recovery: these high-value elements can be effectively recovered using biosorption techniques as an alternative to the conventional unit operations of extractive metallurgy [55].

Today, the validity of biosorption, with all its advantages, is more than demonstrated; the great challenge is to apply this technique to more real situations. In the vast majority of studies, the biosorption process is carried out in distilled water, where the pollutants to be tested are solubilized, and there are few examples where biomass is applied to more real situations [56,57].

3. Biosorption of Metals

Metal biosorption is among the most studied applications of this technique. In fact, the first applications of biosorption focused on metal removal [58,59]. For decades, metals have been a serious environmental problem due to human activity [60–62]. For this reason, the development of techniques that allow their removal has been a priority. Biosorption is a very effective, economical and environmentally friendly technique to remove these pollutants, and at present, different methods have been evaluated based on biosorption [22,63].

Ion exchange is the predominant mechanism for metal biosorption along with surface complexation and microprecipitation [46,55,64]. Various functional groups are involved in the biosorption of metals, carboxyl, hydroxyl, sulfate, phosphoryl and amino groups [31]. Because of this, pH has an important effect on the biosorption of metals. The pH affects the charge of these functional groups and, therefore, the amount of biosorbed metal. Since cationic species are among the predominant forms of most metals in aqueous solution, the more negative charge the biosorbent has, the greater the amount of biosorbed metal. For this reason, the most suitable pH range for metal biosorption is 7.0–8.0. At lower pHs, hydrogen ions and metal ions compete for binding sites; and at higher pHs, there is precipitation of metal ions in the form of hydroxides, reducing the amount of biosorbed metal. However, this behavior is different for the case of some metals whose predominant forms are anionic, such as chromium, arsenic or molybdenum, among others. In this case, acidic pHs (2.0–4.0) are the most favorable for increasing biosorption because at these pHs the biomass has a greater number of positive charges, which allows the attraction of anions.

Although pH is considered a key factor in this process, temperature also influences biosorption since this parameter affects the rate of reactions. Higher temperatures usually enhance biosorption rate due to the increase in surface activity and kinetic energy of sorbate [65]. However, its effect on the maximum amount of biosorbed metal is debatable. It is generally accepted that the increase in temperature increases the maximum amount bioasorbed, which occurs when the process is endothermic, and is due to various factors such as structural changes in the sorbent or breakdown of bonds between the sorbent molecules; however, there is also some exceptions, which take place when the process is exothermic. In this case, there is a decrease in biosorption capacity with an increase in temperature, possibly due to damage caused to the surface of the biosorbent [30,66,67]. With dead biomass, this effect of temperature is less apparent than with living biomass. With living biomass, as the temperature increases, the amount of biosorbed metal increases more appreciably than in the case of dead biomass [68,69]. The reason is the greater metabolic activity of living cells when the temperature increases until an optimum value, causing the metal to be incorporated in a higher amount into the cell interior. When the temperature exceeds the optimum value, the living material is damaged and the biosorption decreases to a greater extent in relation to the dead biomass [27]. Finally, an increase in ionic strength reduces the amount of biosorbed metal due to competition of other cations for the binding sites on the functional groups [70]. This is a major drawback when applying biosorbents in real effluents that are often characterized by complex concentrations of different cations.

Precisely, materials derived from biomass are characterized by offering a large and diverse number of functional groups that interact with metals (carboxyl, hydroxyl, sulfate, phosphoryl and amino groups, as indicated above), which explains the good performance of these materials as biosorbents in metal removal. Of special interest is biomass derived from algae because it has a relatively high adsorption efficiency of 1–10 g/L [71]. As an important additional property, biomass derived from microorganisms can be easily genetically modified to increase the biosorption capacity. This strategy is receiving a lot of attention recently to increase metal removal. Thus, the expression of the EC20 protein (a synthetic phytochelatin) on the surface of various bacteria was used in Pb, Zn, Cu, Cd, Mn, Ni and, recently, in Pt biosorption [34]. In the same way, the expression of a non-MT cadmium-binding protein from *Lentinula edodes* significantly enhanced the cadmium biosorption capacity of transgenic *Escherichia coli* [35]. The transformation of the wild-type *Saccharomyces cerevisiae* with two versions of a *Populus trichocarpa* gene (PtMT2b) coding for a metallothionein allowed an increase in the intracellular content of cadmium in relation to the wild strain [36].

Another aspect to take into account when applying biosorption to metal removal is that biosorption not only serves to remove these elements but also allows their recovery, which increases interest in this process [72]. This application can be extended to an industrial scale, for which the sorbent must have adequate properties. Precisely, one of the interesting properties of biomass is that it can be easily modified to adapt it to commercial and industrial uses [73].

Metal biosorption studies cover most commonly used metals, but metals considered non-essential are the ones that have received the most attention. Numerous studies on metal removal continue to be carried out today using different biosorbents.

3.1. Chromium (VI)

Chromium is the metal that has received the most attention lately for its removal through biosorption (Table 1).

Table 1. Recent chromium (VI) biosorption studies using different biomasses.

Biomass	Chromium Concentration (mg/L)	Contact Time (min)	Efficiency (mg/g)	Remarks	Reference
Waste biomass from polyglutamic acid production	50–400	60	2.39	Immobilized in sodium alginate	[74]
Date palm empty fruit bunch	50–250	120	70.49	Batch studies	[75]
Datura stramonium fruit	100–300	60	138.074	Carbonized form of sulfuric and phosphoric acid-treated biomass	[76]
Halomonas sp. DK4	50–250	2880	150.7	Batch studies	[77]
Zhihengliuella sp. ISTPL4	50–500	1440	49 ± 0.3	Calcite-based biocomposite biochar	[78]
Coconut fibers	100	20	87.38	Magnetized using magnetite nanoparticles	[14]
Quercus crassipes acorn shell	50–400	-	110.35	Fixed-bed column	[79]
Leiotrametes flavida	1000	1200	285	Live and heat-inactivated	[80]
Sargassum horneri	100–600	360	330.84	Polyethyleneimine-modified ultrasonic assisted acid hydrochar	[81]
Pteris vittata	100–250	120	166.7	Dead, unmodified	[82]
Synechococcus mundulus	75–175	2880	- 85.89% - Initial Cr(VI) concentration of 300 mg/L	Extruded polymeric substances secreted under the effect of 2-KGy gamma irradiation dose	[17]
Agaricus campestris	10–100	200	56.21	Unmodified dead biomass	[83]
Pleurotus ostreatus	10–150	22	- Living biomass: 169.84 - Dead biomass: 368.21	Living and dead biomass	[84]
Pseudomonas sp. strain DC-B3	10–135	420	25.64	Living cells	[85]
Kodamaea transpacifica	10–100	240	476.19	Cationic surfactant-modified living cells	[86]
Parapedobacter sp. ISTM3	10–200	15	33.78	Extracellular polymeric substances	[87]
Eupatorium adenophorum	10–300	60	28.011	Calcium alginate entrapped	[26]

As can be seen in this table, the biological materials that have been evaluated are very diverse and show very good efficiency. The strategies using these different biomasses were also very varied since they range from typical batch experiments to continuous flow systems, immobilization techniques or more sophisticated modifications of the biomass, which demonstrates the versatility of biosorption.

The main mechanisms involved in the biosorption of Cr(VI) are related to electrostatic attraction, surface complexation and heterogeneous redox reaction to form Cr(III) ions [75]. In addition, in chromium biosorption processes, it is necessary to consider that this metal, unlike most metals, is in the form of anions. This means that the behavior towards biosorbents is different. In this case, the range of pHs considered optimal to carry out biosorption is 2.0–3.0 [14,26]. At low pHs, the biomass surface is highly protonated, offering a large amount of positive charges that attract chromium anions. Obviously, this pH range cannot be used with living biomass; however, the biological activity of this type of biomass can compensate for this inconvenience, especially using resistant strains [85]. This behavior can be applied to other metals such as dysprosium [88], arsenic [89] or tungsten [20].

3.2. Cadmium (II)

Cadmium is among the metals that has received the most attention from the biosorption field. Today, there are still studies related to this non-essential metal. Thus, living and dead biomass of *Pseudomonas* sp. strain 375 was tested for cadmium removal. Living biomass was more effective (92.59 mg/g) than dead biomass (63.29 mg/g) [8]; it is a strain with great resistance to cadmium toxicity, and for this reason, the living form of this biomass surpassed the dead one in efficiency, demonstrating the interest in testing the use of living systems in this type of applications. Another example of the utility of using living systems was the application of *Pseudomonas chengduensis* strain MBR as living biomass. This strain was able to remove 100% of Cd(II) (with a high initial concentration of 200 mg/L) due to a combination of biosorption and biotransformation. This strain has many functional genes related to heavy metal resistance in its genome which would explain this result [3].

In general, metal tolerant strains show better efficiencies in biosorption of these elements when living biomass is used; for example, living cells of a lead resistant strain of *Staphylococcus aureus* were more efficient for the biosorption of cadmium and lead than dead biomass [44]. This shows that it is very important to screen suitable strains for this purpose [90].

Unlike dead biomass, metabolically active cells can bioaccumulate metal inside the cell, which increases the amount of removed metal. At the same time, this type of biomass can transform the pollutant into non-toxic forms, which is important in practical applications. A similar result was obtained with a cadmium-tolerant bacterium, *Enterobacter ludwigii* LY6: the cadmium chloride removal rate of this strain with a treatment of 100 mg/L of cadmium chloride reached 56.0 %. In this strain, the expression of several genes closely related to bacterial cadmium tolerance and biosorption increased with the increase in the cadmium concentration [91]. Taking this into account, genetic modification is also a very useful tool to achieve resistant strains that can be used as living biomass, and therefore, with better biosorption capacity. Several examples show the effectiveness of this strategy. The deletion of the crpA gene (P-type ATPase) in the fungus *Aspegillus nidulans* showed 2.7 times higher cadmium biosorption capacity [92]. A transgenic yeast that expressed a metallothionein gene from *Populus trichocarpa* had higher intracellular Cd than the wild strain [36]. Through genetic engineering, a plant cadmium and zinc transporter (AtHMA4) was also used as a transgene to increase tolerance to these metals and the biosorption capacity of *Chlamydomonas reinhardtii* [93].

Most studies indicate that the tightly bound cadmium on the cell wall plays a major role in Cd^{2+} adsorption [8,90]. Thus, cadmium biosorption studies with the *Simplicillium chinense* QD10 strain [94] and with *Shewanella putrefaciens* [95] used as living cells suggested that the cell wall components were the primary interactive targets for this metal. Cadmium sulfide nanoparticles can also form on the cell surface, which contributes to the excellent tolerance to this metal of *E. ludwigii* LY6 [91]. Additionally, on the cell surface, the exopolysaccharides (EPS) might be the main means of cadmium adsorption by some strains [91].

Another recently used approach to increase the efficiency of cadmium removal was the use of grapefruit (*Citrus paradisi*) peel treated with Ca^{2+} or Mg^{2+}. Through these modifications, increases of 46.3 and 27%, respectively, were achieved in the amount of cadmium removed by this biomass, demonstrating that this residue with a simple modification can be useful as a cadmium biosorbent [96]. A novel composite, which was synthesized by *Bacillus* sp. K1 loaded onto Fe_3O_4 biochars, presented a 230% increase in the capacity to remove cadmium compared to raw magnetic biochar [89]. This is one more demonstration of the importance of materials of biological origin in sorption processes.

The *Pediococcus pentosaceus* FB145 and FB181 strains, which can be considered as probiotic microorganisms, were suggested as potent biosorbents for preventing cadmium toxicity and reducing its absorption into the human body [97]—one more utility of biosorption, in this case, directly related to human health.

3.3. Lead (II)

Lead is another non-essential element that recently attracted greater attention from the field of biosorption. A new proposal to improve the biosorption of this element by fungal biomass was developed using the biomass of *Phanerochaete chrysosporium* with an intracellular mineral scaffold. The intracellular mineral scaffold of this functionalized biomass served as an internal metal container exhibiting high biosorption efficiency for Pb(II) and Cd(II) ions [98]. A comparative study using different biomass of microorganisms (*Pseudomonas putida* I3, *Microbacterium* sp. OLJ1 and *Talaromyces amestolkiae*) showed that the different cell structure had a clear influence on the efficiency of lead biosorption. The most efficient biomass was *Pseudomonas putida* I3 with 345.02 mg/g. These biosorbents were tested in real wastewater, revealing that these biosorbents possessed good environmental adaptability and great potential for the removal of trace heavy metals from wastewater in practical application [46].

Live and dead biomass of a highly Pb(II) resistant (up to 2200 mg/L) bacterium (*Bacillus xiamenensis*) were also tested for biosorption of this ion. The maximum Pb(II) uptake was 216.75 and 207.4 mg/g for live and dead biomass, respectively [99]. Again, the living and active biomass of a resistant strain showed better performance (intracellular accumulation of lead ions was detected). Living and dead biomass of a (Pb)-resistant bacterium, *Staphylococcus hominis* strain AMB-2 was also evaluated for lead and cadmium removal. Living biomass exhibited more biosorption of metals than dead biomass in both single and binary systems; moreover, lead had a higher affinity for the binding sites on the biomass surface [44]. However, a different result was obtained using living and dead biomass of *Rhodococcus* sp. HX-2. In this case, the dead biomass was more effective. The maximum biosorption capacities were 88.74 and 125.5 mg/g for live and dead biosorbents, respectively. In this case, Pb(II) adhered to the surface of dead biosorbents more easily than to the surface of live biosorbents [68]. The characteristics of the cell surface also affect the amount of lead removed because, in addition to biosorption, this ion can be mineralized [100].

Biomass from the *Simplicillium chinense* fungus strain QD10 had a maximum biosorption capacity of 57.8 mg/g. In this strain, the lead biosorption was predominantly adsorbed by extracellular polymeric substances [101].

Other biomasses that were also recently tested for lead biosorption were *Moringa oleifera* leaves (maximum biosorption of 45.83 mg/g) [102] and the lactic acid bacterium *Lactobacillus brevis* used as living biomass with a maximum biosorption of 53.63 mg/g [69]. Cotton (*Gossypium hirsutum*) shell powder was used as a biosorbent for the treatment of synthetic Pb-contaminated water. This biomass reached a biosorption capacity of 9.6 mg/g [103].

3.4. Mercury (II)

Mercury is another of the non-essential elements that recently had some study from the point of view of biosorption. Different tests show the application of this technique to remove this metal. A biopolymer consisted of proteins, carbohydrates and nucleic acids from waste activated sludge was evaluated. This biopolymer had a maximum adsorption capacity up to 477.0 mg Hg(II)/g [21]. Algal biomass (*Chlorella vulgaris* UTEX 2714) was also tested for remove Hg(II). This biomass used as dead biomass presented a rapid kinetics of adsorption (90 min) and with a capacity of 42 mg/g [104]. In addition, this biomass presented a good regeneration: this property is important for a biosorption process to be viable.

Living systems were also recently used to remove mercury. Living biofilm, developed on a non-woven polypropylene and polyethylene geotextile was tested. This biomass removed 13.34 mg/g in 28 days [105].

3.5. Uranium (VI)

Uranium is another element that has recently received attention from the field of biosorption. This metal is a major health problem; therefore, the development of applications for its removal shows considerable interest. Different biomasses have recently been successfully evaluated for this purpose; of these biomasses stands out *Saccharomyces cerevisiae*. Dead biomass of this yeast removed uranium efficiently due to the large number of functional groups that this biomass presents [31]. This same species, immobilized by a new method based on saturated boric acid-alginate calcium cross-linking, had a biosorption capacity of 113.4 μmol/g [106].

Living biomass of the resistant bacterium, *Bacillus amyloliquefaciens* had a maximum uptake capacity of 179.5 mg/g [45]. Similarly, biomass from the macrophytes *Pistia stratiotes* and *Lemna* sp. presented a maximum uranium sorption capacity of 2.86×10^{-2} and 6.81×10^{-1} mmol/g, respectively, with an optimum contact time of 60 min [107].

Sorbent modification has been a widely used method to increase the efficiency of uranium removal. This method is also applicable when biomass is used. Tri-amidoxime modified marine fungus material had a uranium biosorption capacity of 584.60 mg/g with good regeneration performance. The unmodified biomass originally had a capacity of only 15.46 mg/g [108]. The macroporous ion-imprinted chitosan foams showed an adsorption capacity of 248.9–253.6 mg/g. This modification of chitosan increased the number of active sites, mainly amine and hydroxyl groups, increasing the coordination with U(VI) [109].

Including uranium, the biosorption technique has recently been shown to be useful for removing metals from radioactive liquid organic waste. Rice and coffee husks (raw and chemically activated) were examined regarding their capacity to remove U(total), ^{241}Am and ^{137}Cs, demonstrating that these materials can be used for the treatment of this waste [110].

3.6. Copper (II)

Despite the fact that copper is an essential element, when its concentration is high, it is potentially toxic. This generates the need to develop procedures for its elimination from natural environments. Biosorption proves to be a very useful tool for this purpose. Several biosorbents have recently been evaluated to remove this metal. Biomass from the ornamental herb *Thevetia peruviana* had a biosorption capacity of 187.51 mg/g, far superior to other biomass or pretreated materials [19]. A new *Alcanivorax* sp. VBW004, resistant to copper toxicity, isolated from the shallow hydrothermal vent (Azores, Portugal) was evaluated for biosorption of this metal. This live biomass, cultured with 100 mg/mL of copper, reduced the concentration of this metal by 39.5 % after 48 h. Genetic studies revealed that this strain has copper detoxification genes [111]. This fact shows once again that living biomass with adequate characteristics can be superior to dead biomass.

Immobilization was also recently used for copper biosorption. Alginate-immobilized cells (living biomass) of the bacteria *Azotobacter nigricans* NEWG-1 was able to remove a percentage of copper of 82.35 ± 2.81% after 6 h and with an initial copper concentration of 200 mg/L [112]. The biomass of the *Aspergillus australensis* fungus was also used in immobilized form. In this case, commercial samples of a textile made of 100% polyester were used as an immobilization matrix, living and dead biomass were compared. In this study, it was observed that an active biosorption process took place, resulting in a higher copper removal compared to a passive process [11]. Another example of an immobilized system is the use of biomass from sugar beet shreds in a fixed-bed column. This process was optimized to remove copper using Box-Behnken experimental design with concentration and pH of the inlet solution and adsorbent dosage as independent variables [24].

Other biomasses that were also tested to assess copper biosorption were *Chlorella pyrenoidosa*, reaching 0.48 mmol/g [41] and *Ochrobactrum* MT180101: in this strain, there were several mechanisms involved in the biosorption of this metal: surface biosorption, extracellular chelation and bienzyme-mediated biotransformation, which supposes a superior efficiency in the copper biosorption [5]. The commercial biomass of the yeast *Saccharomyces cerevisiae* Perlage® BB with a maximum biosorption capacity of 4.73 mg/g [113], dead biomass of *Penicillium ochrochloron* with an average biosorption capacity of 7.53 mg/g [114], *Sargassum filipendula* [42] and alginate-based biosorbent produced from seaweed *Sargassum* sp. with a maximum biosorption capacity of 1.64 mmol/g [70] are recent examples of different biomasses that have been evaluated to determine their capacity as copper biosorbents.

3.7. Other Metals

Other metals that have recently been studied from the point of view of biosorption are shown in Table 2.

Table 2. Examples of other metals that have recently been studied for their removal through applications related to biosorption.

Metal	Biomass	Remarks	Reference
Zinc	Leaves of *Corchorus olitorius*	Biosorption capacity of 11.63 mg/g in 120 min	[115]
	Streptomyces K11	- Living biomass - Maximum biosorption capacity of 0.75 mmol/g	[65]
	Rape straw powders	- Different parts were evaluated - The maximum biosorption capacity obtained reached 36.74 mg/g	[116]
Antimony	*Rhodotorula mucilaginosa* Strain DJHN070401	- Living cells - Providing that living cells not only improved the removal efficiency in the presence of metabolic inhibitors but also prevented intracellular Sb(III) being re-released into the environment	[9]
Gold	*Lysinibacillus sphaericus* CBAM5	- Biocomposite materials - Cells immobilized in Polycaprolactone (PCL) microfibrous mats and alginate microcapsules	[28]
Arsenic	*Sarcodia suiaeand*	External factors affecting algal metabolism and thus metal-accumulation mechanisms were studied	[67]
	Pseudomonas aeruginosa AT-01 Strain	Remediation efficiency at 2 h of incubation was 97.92 % with an initial As concentration of 10 mg/L	[117]
Manganese	*Providencia* sp. LLDRA6	- Mn(II) precipitation on the cell surface - Oxidation of Mn(II) into BioMnOx on the cell surface - Intracellular accumulation	[118]
Platinum	*Escherichia coli*	- EC20 protein fused to the *E. coli* cell surface using an InaKN-based display - Maximum adsorption capacity of 239.92 mg/g	[34]
	Lemon peel	Maximum adsorption capacity of 36.74 mg/g	[33]
Nickel	*Phanerochaete chrysosporium*	- Living cells - Maximum biosorption capacity of 46.50 mg/g	[119]
	Surface-engineered *Saccharomyces cerevisiae* EBY100	Biosorption capacity of 2.603 ± 0.004 g/g	[120]
	Industrial waste brewery sludge	Biosorption capacity of 7.874 mg/g	[15]
	Alginate-based biosorbent produced from *Sargassum* sp.	Maximum biosorption capacity of 1.147 mmol/g	[70]
	Fucus vesiculosus	- Simultaneous biosorption of Cd(II), Ni(II) and Pb(II) - Maximum biosorption capacity of 70.1 mg/g for Ni(II)	[66]
Lithium	*Aspergillus versicolor* and *Kluyveromyces marxianus*	Maximum biosorption capacities of 347.9 and 409.2 µmol/g for *A. versicolor* and *K. marxianus*, respectively	[47]
Cesium	*Haematococcus pluvialis* and *Chlorella vulgaris*	Cesium accumulation through the potassium transport channel	[121]
Dysprosium	*Mangifera indica*	- Raw and surface-modified bark powder - Maximum adsorption capacity of 55.04 mg/g	[88]
Tungstate	Garlic peel	- Modified by loading with Fe(III), Ti(IV) and Ce(III) through a cation exchange process - Maximum adsorption capacity of 91.5 mg/g with Fe	[20]
Lanthanum and samarium	*Botryosphaeria rhodina* MAMB-05	- Living and dead biomass	[122]
Iron	Elderberry (*Sambucus nigra*) pomace	- Fe (III) - Maximum biosorption capacity of 33.25 mg/g	[123]
	Bacillus subtilis	- Fe(II) - Living cells - Maximum biosorption capacity of 7.25 mg/g	[124]

4. Biosorption of Organic Compounds

Today, many organic compounds produce undesired effects in natural ecosystems, and some are considered very toxic to humans. Many of these are part of the so-called persistent organic pollutants (POPs) such as pesticides, insecticides, organochlorines, herbicides and polychlorinated biphenyls (PCBs). Although many of these compounds have been known and used for a long time, some are of recent development, and others have been discovered in the environment due to the progress of analytical techniques: these compounds have been called emerging organic contaminants (EOCs) [125,126]. Because these compounds cause serious problems in ecosystems even at low concentrations, it is necessary to develop techniques for their elimination. Physico-chemical techniques are sometimes not effective, they are more expensive and they can also generate additional problems.

In this context, biosorption is an alternative that avoids these inconveniences and for this reason is being increasingly developed to remove this type of substance. This is demonstrated by the number of studies that have recently been carried out to use biosorption in the elimination of these compounds.

The process is performed in a similar way to metal biosorption and, in general, the factors affecting efficiency are the same, although the response to them shows differences that must be studied in each case. Unlike metals, the complexity of these compounds, in terms of their composition, means that they may have different functional groups capable of presenting very different charge values and with different degrees of ionization depending on the pH of the solution. For this reason, the optimization of this parameter is of great importance, and the optimization values obtained for compounds with different nature show greater diversity than in the case of metals.

Furthermore, these compounds may have different degrees of hydrophobicity and reactivity, which have an effect on the process. Although hydrophobic compounds are not readily soluble in water, such compounds can interact with the biosorbent particles through hydrobophic interaction or can even cross the cell membranes when using living biomass. Therefore, this type of compounds can also be removed by biosorption.

In relation to temperature, its effect is contradictory, and generally the adsorption effectiveness increases with increasing temperature (endothermic process) [56,127]. However, there are results with some organic compounds whose effect was the opposite, indicating in these cases that the biosorption process was exothermic [128]. Finally, the ionic strength of the solution also modifies the biosorption capacity of organic compounds, although its effect seems to be less relevant than in the case of metals. A high concentration of salts is necessary for a significant decrease in the biosorption of these compounds to occur, although this effect is less studied than in the case of metals.

Some examples of organic compounds that have recently been studied will be reviewed in the following sections.

4.1. Antibiotics

Antibiotics are found today in relatively high amounts in ecosystems due to their increasing use. In fact, many studies have been carried out to examine the possibilities of biosorption for the removal of antibiotics. Recently, new articles confirm this use. An example of this is the antibiotic dicloxacillin, biosorption studies with this antibiotic that were performed with Indian almond (*Terminalia catappa*) leaf biomass. The maximum adsorption capacity was 71.94 mg/g. The optimal pH for this biosorption was 6.0. For the intermolecular interaction such as hydrogen bonds, van der Waals forces could be the main interaction for the dicloxacillin and the surface properties of this biosorbent [129]. Four sulfonamide antibiotics were also recently investigated. Extracellular polymeric substances (EPS) extracted from *Klebsiella* sp. J1 were used for this purpose and with efficiencies that reached 142.86 mg/g. In this case, the hydrophobic interaction between EPS and sulfonamides seemed to dominate the adsorption process. There was the maximum proportion of sulfonamides at molecular states when the solution pH was 4.0–5.0, and the molecular states of sulfonamides were favorable for hydrophobic groups to effectively play a role during adsorption process [130].

Biomass from *Dialium guineense* seed waste was modified with sodium hydroxide and tested as biosorbent for ciprofloxacin. This biomass, modified in this way, exhibited a maximum uptake capacity of 120.34 mg/g at pH = 6.0, higher than some reported adsorbents for this antibiotic [131]. Another antibiotic that was recently evaluated for its elimination by biosorption was oxytetracycline. This antibiotic was effectively removed using a reed-based-beads biosorbent (an enhanced adsorbent from Tunisian reed). The maximum biosorption capacity obtained was 15.78 mg/g at pH = 6.0. In this experiment, an initial mixture of 165.54 μmol/L of oxytetracycline and 362.16 μmol/L of Cd(II) was used, demonstrating the effectiveness of this material to simultaneously remove two pollutants. The biosorption of Cd(II) cations took place through electrostatic attraction between them and the biosorbent, and the biosorption of oxytetracycline could take place via π-π stacking, as well as hydrophobic interactions [52].

4.2. Dyes

Today, dyes are one of the main pollutants. They are produced in large quantities due to their widespread use, and for this reason, they can be released into natural environments. The release of these types of compounds causes serious problems to natural ecosystems due to their toxicity, carcinogenicity and because they also impart an intense color to the waters (even at very low concentrations) and, in addition, they are considered recalcitrant compounds. Therefore, effective treatments are necessary to remove these compounds from the waters. Biosorption is an excellent alternative to conventional methods. In fact, many papers have been published on the removal of these pollutants by biosorption, and recently, dyes continue to dominate biosorption applications.

The cationic dye methylene blue is one of the most widely used dyes in the industry, and for this reason, there is a wide variety of biosorbents that have been evaluated for the elimination of this dye. In recent years, there are also various examples of these proposals, in which both modified and unmodified biomass are used. Thus, coconut waste, chemically modified with acrylic and polyacrylic acids, has been utilized for the removal of this dye. With these modifications, the maximum sorption capacity reached was 138.88 mg/g at pH = 10.0 using the acrylic acid. The sorption mechanism is mainly based on electrostatic interaction and on Lewis acid-base interaction [30]. *Cortaderia selloana* flower spikes transformed into nanomagnetic particles reached a maximum removal capacity of this dye of 119.05 mg/g at pH = 6.0 and through electrostatic interactions [132].

However, unmodified biomass methods are still the most widely used to remove methylene blue. Recently, it has been made possible to find many examples of this. Brewer's spent grain has recently been used with a maximum adsorbed amount of methylene blue of 284.75 mg/g at pH = 11.0. A possible adsorption mechanism involves electrostatic interaction, electron donors and electron acceptors, hydrogen bonds and π-π dispersion [133]. Biomass obtained from weeds (*Cyanthilium cinereum* and *Paspalum maritimum*) was also evaluated to eliminate this dye, the maximum absorption capacities obtained were 56.18 and 76.34 mg/g, respectively [134]. *Fucus vesiculosus* dead biomass was also used to remove this dye, presenting a maximum biosorption capacity of 698.48 mg/g at pH = 6.0 by a physical biosorption mechanism related to a cation exchange process between the dye and biomass functional groups, releasing protons (H^+) to the system [135]. Biomass from *Bifurcaria bifurcata* was also applied to remove this dye with a maximum biosorption capacity of 2744.5 mg/g in only 15 min. The best adsorption efficiency was obtained at pH = 5.6 due to electrostatic interaction [136]. In these examples, although the biomasses of these macroalgae were clearly higher, it is necessary to consider that the brewer's spent grain is a byproduct of the brewing industry produced in large quantities and with few ecological disposal options; in addition, weeds are abundant and with few applications. This shows that the choice of a certain biosorbent depends on many factors that must be evaluated together. The search for new biosorbents is essential to achieve this objective. An interesting example of this is the application of the biomass from brazilian berry seeds (*Eugenia uniflora*) to remove this dye in conditions closer to reality using two different simulated effluents with a color removal greater than 70%. This material had a biosorption capacity of 189.6 mg/g at pH = 8.0 and presents OH groups that can perform hydrogen and electrostatic bonds with methylene blue [56]. Other macroalgae also recently used as biomass were *Ulva fasciata* that reached a maximum adsorption capacity of 244 mg/g and *Sargassum dentifolium* with only 66.6 mg/g. In both cases, it was determined that the biosorption of methylene blue was independent of pH [12].

Crystal violet is another cationic dye that has recently received attention for biosorption. Powdered seeds of the araticum fruit (*Annona crassiflora*) were used in the biosorption of this dye with a maximum biosorption capacity of 300.96 mg/g in 120 min at pH = 7.5. Electrostatic interaction played an important role in the biosorption process of this dye since a pH higher than the zero charge point of the biosorbent (7.2) causes the surface of this material to be negatively charged due to its deprotonation, which favored an increase in its biosorption capacity because crystal violet is a cationic dye, and therefore, interaction with a negative charge on the surface increases the amount of dye biosorbed [18]. Biomass from *Diaporthe schini* (new fungus recently discovered) reached a maximum biosorption capacity of 642.3 mg/g. This biomass was also evaluated in a simulated effluent with a considerable reduction in the color [137]. Finally, as a more sophisticated option, water dispersible Fe_3O_4/Chitosan/Glutaraldehyde nanocomposites (superparamagnetic) were also tested, in this case with a maximum biosorption capacity of 105.47 mg/g. Electrostatic interaction between the negative charge of the biosorbent surface and the positive charge of the dye would explain why this maximum capacity took place at pH 11.0. Although the maximum removal capacity was lower than the previous cases, the advantage of this technique is its ease and rapid separation from samples, allowing a reusability up to at least ten cycles [138].

The discarded seed biomass from pepper (*Capsicum annuum*) was tested to remove Basic red 46 dye. Taguchi DoE methodology was employed to optimize the process reaching a dye removal performance of 92.1 mg/g at pH = 8.0 [139].

Biomass derived from macroalgae was also evaluated to remove the Rhodamine B dye. The macroalgae used were *Kappaphycus alvarezii*, *Gracilaria salicornia* and *Gracilaria edulis*, both in native form and ethanol modified. The maximum biosorption capacity determined as 9.84, 11.03, 8.96, 112.35, 105.26 and 97.08 mg/g at pH = 2.0, respectively. At this low pH, there is an increase in the protonation effect on the surface of these materials resulting in a higher biosorption capacity. The modified biomass was more efficient [140].

Anionic dyes are the other group of dyes that also have a multitude of applications, and for this reason, they are also an environmental problem. These types of dyes have also recently been studied as applications in the field of biosorption. Thus, the removal of tartrazine yellow was evaluated using brewer's spent grain as biomass; the maximum adsorbed amount was 26.18 mg/g at pH = 2.0. This adsorption involves electrostatic attraction, π-π interaction and multilayer formation of dye [133]. Reactive Blue 19 using dead biomass of the brown marine alga *Bifurcaria bifurcata*, with a maximum adsorbed amount of 88.7 mg/g in only 15 min at pH = 1.0. At this very low pH, the concentration of H_3O^+ was high enough to allow the protonation of sulfonate groups of this dye, which favors the interaction between the dye and the functional groups of the biomass [136]. Eriochrome black T is another example of an anionic dye recently studied for its elimination by biosorption; in this case, using dead biomass of *Fucus vesiculosus* and with a maximum biosorption capacity of 24.31 mg/g at pH < 4.0. Van der Waals interaction was the main interaction mechanism between this dye and biomass [135]. Biomass from *Ocimum gratissimum* leaves was tested for the indigo carmine dye biosorption. This biomass obtained a maximum biosorption capacity of 77.52 mg/g, confirming that this capacity was superior to that of other sorbents used to remove this dye. Since this dye is anionic, the most favorable adsorption occurred at pH 2.0 because the surface of this biosorbent is positively charged [128]. Direct Fast Scarlet 4BS was successfully removed using dead biomass from *Enteromorpha prolifera* with a maximum sorption capacity of 318.87 mg/g also at pH 2.0. The adsorption mechanism involved hydrogen bonding, electrostatic attraction and bonding and hydrophobic and van der Waals interaction [127]. Finally, Reactive Red 120, using immobilized biomass of *Pseudomonas guariconensis* in a Ca-Ag biocarrier matrix, was efficiently eliminated. In this case, since the biomass was alive, in addition to biosorption, biodegradation occurred. Toxic reactive dye was converted into non-toxic compounds. The immobilized bacterial cells exhibited 87% uptake of this dye, whereas the non-immobilized bacterial cells exhibited a maximum uptake of 37% [6].

Anionic dyes are better adsorbed at low pHs—that is, at pHs below the zero-charge point of the biosorbent—because under this condition, the surface of the biosorbent acquires positive charge.

4.3. Other Organic Pollutants

Other organic pollutants of interest have also been treated using biosorption as the primary removal technique. Table 3 shows some examples of the most recently studied organic pollutants.

Table 3. Examples of other organic pollutants that have recently been evaluated for their removal using biosorption.

Pollutant	Biomass	Efficiency	Remarks	Reference
Phenol	*Luffa cylindrica*	28.9 mg/g	Hybrid material with 4% Zn^{2+}	[38]
Sterols	*Aspergillus fumigatus* strain LSD-1	303.03–909.09 mg/g	Living and dead biomass	[141]
Propranolol hydrochloride	*Sargassum filipendula*	1.94 mmol/g	Remaining biomass of alginate extraction	[142]
Acetylsalicylic acid	Biosorption onto fungal-bacterial biofilm supported on two types of activated carbons	292.4 ± 2.01 mg/g	Batch and fixed-bed experiments	[143]
Nonylphenol	Microalgae	74.18–92.12% in 120h with initial concentration 1 mg/L	Living biomass	[7]
17 alpha-ethinylestradiol alone and along with estrone	Yeast biomass from ethanol industry	24.50 ± 0.07 and 0.80 ± 0.07 mg/g	An associative/competitive sorption process between both compounds	[40]
Salicylic acid	*Scenedesmus obliquus*	63 mg/g	- Batch experiments - Dead biomass	[144]
Diuron	*Moringa oleifera*	5.76 mg/g	Fixed-bed column	[25]
Ibuprofen	*Scenedesmus obliquus*	11.9 mg/g	- Batch experiments - Dead biomass	[144]
Triclosan	*Phaeodactylum tricornutum*	12.97–13.03 mg/g	- Seawater - Living and dead biomass and photodegradation	[10]

As can be seen in this table, the nature of the organic compounds is very varied, which is indicative of the enormous possibilities that biosorption techniques have for the removal of this type of pollutant. In addition, these examples reflect the flexibility of biosorption techniques since, in the same way as for other pollutants, biomass can be alive or dead, in batch or in fixed-bed experiments, but it is noteworthy that even this technique can be coupled to an alternating current system that allows increasing the speed of biosorption [38]. However, the removal capacity that some of these sorbents have is far from that achieved with commercial sorbents such as activated carbon. Although the comparison data are scarce, the values obtained indicate the need to search for biomaterials with greater capacity—for example, the biomass of *Scenedesmus obliquus* had a maximum removal capacity of salicylic acid and ibuprofen of 63 and 11.9 mg/g, respectively; instead in the same conditions, activated carbon had 250 and 147 mg/g [144].

5. Conclusions and Future Perspectives

As can be seen, the field of biosorption continues to offer very promising results for the elimination of pollutants. It is a technology that presents a great diversity of options and combinations, demonstrating great flexibility for its application. It is difficult to limit the studies that try to reveal the properties of any material for use in biosorption. The reason for this is obvious: the amount of possible materials (living or dead) is enormous. These studies must continue to progress because without a material with adequate properties, biosorption cannot be competitive. However, it is not only necessary to determine the properties of a possible material but also to evaluate it compared to others already established as sorbents (commercial sorbents), and therefore, conclude that this new material is a better alternative. There are still steps to be taken for biosorbents to be fully accepted. Currently, there are several challenges of biosorption: the development of large-scale procedures, greater commercialization and, in general, its application in real conditions. Although the advantages of this type of sorbent are evident (mainly cost), few biosorbents are currently marketed for their use [22]. The application of biosorption at the industrial scale has not been yet well exploited, and this constitutes another of the weaknesses that biosorption must face. Still, the vast majority of biosorption applications focus on laboratory studies. All these studies make possible the current knowledge about biosorption that is enough to provide a solid base that allows its use to be extended. However, this process is not widely used in industry.

An important reason that can explain these weaknesses is that biosorbents, in their natural state, tend to have a lower removal capacity than traditional or conventional sorbents such as activated carbons, zeolites or ion exchange resins. However, it is difficult to ascertain this fact because at present there are still few studies in which a biosorbent is compared with commercial sorbents under the same conditions [145,146]. Perhaps, the fact of thinking that a biosorbent may be less effective than the traditional ones could be counterproductive to achieve that biosorbents climb positions, because the true capacity of these materials will remain unknown. The fact that the matrices used in the experiments have different physicochemical properties does not help much in improving the perception of biosorption, since this makes it difficult to compare biosorbents to obtain the one with the highest affinity for a pollutant. A certain standardization could be interesting to solve this aspect.

In any case, if the above is true and biosorbents lack the necessary efficiency, this would imply the need to modify them to achieve greater efficiency. It would be desirable if the biosorbents had, at least, characteristics comparable in efficiency to the commercial ones. There are several alternatives that can improve the effectiveness of biosorbents, ranging from chemical or physical modifications to the use of nanomaterials [29,63,73]. Chemical of physical modifications applies mainly to dead biomass. However, these alternatives would increase the cost of the final product, and the resulting material could be even less eco-friendly, reducing the virtues of biomaterials. In this context, it is interesting not to forget the use of living biomass to improve the effectiveness of a pollutant removal process. Despite the advantages attributed to dead biomass, the properties of living biomass for application as biosorbents have not yet been adequately exploited or even better studied. Many studies indicate that the use of living biomass is more efficient than dead biomass, and living biomass is used without modification. The cost of the production and maintenance of living biomass is among the problems attributed to the use of this type of biomass; however, there are organisms that can be cultivated intensively and with low cost [147]. Macrophytes, microorganisms such as microalgae or some species of bacteria offer very promising results. Studies in this direction should continue.

Immobilization is another key mechanism to improve biosorption processes. It is also a fundamental mechanism for the application of biosorbents on an industrial scale. At present, different proposals are still being evaluated to solve the practical problems of immobilization, especially when living biomass is used [106]. In fact, there has been an increase in the number of studies using immobilized living biomass, perhaps because for many industrial applications, the use of living biomass is preferable. The support for this biomass is being increasingly perfected, as well as the search for the most suitable living biomass for each case, which is essential to ensure that a biosorbent can be successful in its application. An example of this improvement is an alternative that is being exploited by combining nanoparticles with biomass. Microbial cells immobilized on magnetic nanoparticles is a relevant new technique applied to obtain new biosorbents, which has several advantages [29]. However, it is necessary to recognize that the cost of these biocomposite materials can be uncompetitive, as well as an option that can be considered not very eco-friendly. In any case, living biomass immobilization techniques must continue to be refined, seeking more natural and cheaper supports.

Another weakness of biosorption is that many biosorption studies use synthetic wastewater or solutions in distilled water, which does not take into account the behavior of these sorbents with different competitors or with physicochemical parameters that can differ considerably in real conditions. The evaluation of biosorbents in real situations would provide more information and would allow a more adequate assessment of the possibilities of this technique. For this reason, future research should be directed in this direction.

Finally, it is hoped that in the future, as the weaknesses are resolved, biosorption will find its place in industry and in separation technologies.

Funding: This work was carried out with the financial support of the Spanish "Ministerio de Economía, Industria y Competitividad" (CTM2017-88668-R).

Conflicts of Interest: The authors declare no conflict of interest.

References

1. Santaeufemia, S.; Torres, E.; Mera, R.; Abalde, J. Bioremediation of oxytetracycline in seawater by living and dead biomass of the microalga *Phaeodactylum tricornutum*. *J. Hazard. Mater.* **2016**, *320*, 315–325. [CrossRef] [PubMed]
2. Yu, R.; Chai, H.; Yu, Z.; Wu, X.; Liu, Y.; Shen, L.; Li, J.; Ye, J.; Liu, D.; Ma, T.; et al. Behavior and Mechanism of Cesium Biosorption from Aqueous Solution by Living *Synechococcus* PCC7002. *Microorganisms* **2020**, *8*, 491. [CrossRef] [PubMed]
3. Wang, X.; Li, D.; Gao, P.; Gu, W.; He, X.; Yang, W.; Tang, W. Analysis of biosorption and biotransformation mechanism of *Pseudomonas chengduensis* strain MBR under Cd(II) stress from genomic perspective. *Ecotoxicol. Environ. Saf.* **2020**, *198*, 110655. [CrossRef] [PubMed]
4. Rogowska, A.; Pomastowski, P.; Rafinska, K.; Railean-Plugaru, V.; Zloch, M.; Walczak, J.; Buszewski, B. A study of zearalenone biosorption and metabolisation by prokaryotic and eukaryotic cells. *Toxicon* **2019**, *169*, 81–90. [CrossRef] [PubMed]
5. Peng, H.; Li, D.; Ye, J.; Xu, H.; Xie, W.; Zhang, Y.; Wu, M.; Xu, L.; Liang, Y.; Liu, W. Biosorption behavior of the *Ochrobactrum* MT180101 on ionic copper and chelate copper. *J. Environ. Manag.* **2019**, *235*, 224–230. [CrossRef] [PubMed]
6. Reddy, S.; Osborne, J.W. Biodegradation and biosorption of Reactive Red 120 dye by immobilized *Pseudomonas guariconensis*: Kinetic and toxicity study. *Water Environ. Res.* **2020**. [CrossRef]
7. Wang, L.; Xiao, H.; He, N.; Sun, D.; Duan, S. Biosorption and Biodegradation of the Environmental Hormone Nonylphenol by Four Marine Microalgae. *Sci. Rep.* **2019**, *9*, 5277. [CrossRef]
8. Xu, S.; Xing, Y.; Liu, S.; Hao, X.; Chen, W.; Huang, Q. Characterization of Cd(2+) biosorption by *Pseudomonas* sp. strain 375, a novel biosorbent isolated from soil polluted with heavy metals in Southern China. *Chemosphere* **2020**, *240*, 124893. [CrossRef]
9. Jin, C.S.; Deng, R.J.; Ren, B.Z.; Hou, B.L.; Hursthouse, A.S. Enhanced Biosorption of Sb(III) onto Living *Rhodotorula mucilaginosa* Strain DJHN070401: Optimization and Mechanism. *Curr. Microbiol.* **2020**, *77*, 2071–2083. [CrossRef]
10. Santaeufemia, S.; Abalde, J.; Torres, E. Eco-friendly rapid removal of triclosan from seawater using biomass of a microalgal species: Kinetic and equilibrium studies. *J. Hazard. Mater.* **2019**, *369*, 674–683. [CrossRef]
11. Contreras-Cortes, A.G.; Almendariz-Tapia, F.J.; Cortez-Rocha, M.O.; Burgos-Hernandez, A.; Rosas-Burgos, E.C.; Rodriguez-Felix, F.; Gomez-Alvarez, A.; Quevedo-Lopez, M.A.; Plascencia-Jatomea, M. Biosorption of copper by immobilized biomass of *Aspergillus australensis*. Effect of metal on the viability, cellular components, polyhydroxyalkanoates production, and oxidative stress. *Environ. Sci. Pollut. Res. Int.* **2020**, *27*, 28545–28560. [CrossRef]
12. Moghazy, R.M.; Labena, A.; Husien, S. Eco-friendly complementary biosorption process of methylene blue using micro-sized dried biosorbents of two macro-algal species (*Ulva fasciata* and *Sargassum dentifolium*): Full factorial design, equilibrium, and kinetic studies. *Int. J. Biol. Macromol.* **2019**, *134*, 330–343. [CrossRef]
13. Jaafari, J.; Yaghmaeian, K. Optimization of heavy metal biosorption onto freshwater algae (*Chlorella coloniales*) using response surface methodology (RSM). *Chemosphere* **2019**, *217*, 447–455. [CrossRef] [PubMed]
14. Carvalho Costa, A.W.M.; Guerhardt, F.; Ribeiro Junior, S.E.R.; Canovas, G.; Vanale, R.M.; de Freitas Coelho, D.; Ehrhardt, D.D.; Rosa, J.M.; BasileTambourgi, E.; Curvelo Santana, J.C.; et al. Biosorption of Cr(VI) using coconut fibers from agro-industrial waste magnetized using magnetite nanoparticles. *Environ. Technol.* **2020**, 1–12. [CrossRef] [PubMed]
15. Kulkarni, R.M.; Vidya Shetty, K.; Srinikethan, G. Kinetic and equilibrium modeling of biosorption of nickel (II) and cadmium (II) on brewery sludge. *Water Sci. Technol.* **2019**, *79*, 888–894. [CrossRef] [PubMed]
16. Taki, K.; Gogoi, A.; Mazumder, P.; Bhattacharya, S.S.; Kumar, M. Efficacy of vermitechnology integration with Upflow Anaerobic Sludge Blanket (UASB) and activated sludge for metal stabilization: A compliance study on fractionation and biosorption. *J. Environ. Manag.* **2019**, *236*, 603–612. [CrossRef]
17. Hussein, M.H.; Hamouda, R.A.; Elhadary, A.M.A.; Abuelmagd, M.A.; Ali, S.; Rizwan, M. Characterization and chromium biosorption potential of extruded polymeric substances from *Synechococcus mundulus* induced by acute dose of gamma irradiation. *Environ. Sci Pollut. Res. Int.* **2019**, *26*, 31998–32012. [CrossRef]

18. Franco, D.S.P.; Georgin, J.; Drumm, F.C.; Netto, M.S.; Allasia, D.; Oliveira, M.L.S.; Dotto, G.L. Araticum (*Annona crassiflora*) seed powder (ASP) for the treatment of colored effluents by biosorption. *Environ. Sci. Pollut. Res. Int.* **2020**, *27*, 11184–11194. [CrossRef]

19. Medhi, H.; Chowdhury, P.R.; Baruah, P.D.; Bhattacharyya, K.G. Kinetics of Aqueous Cu(II) Biosorption onto *Thevetia peruviana* Leaf Powder. *ACS Omega* **2020**, *5*, 13489–13502. [CrossRef]

20. Wang, Y.; Huang, K. Biosorption of tungstate onto garlic peel loaded with Fe(III), Ce(III), and Ti(IV). *Environ. Sci. Pollut. Res. Int.* **2020**, *27*, 33692–33702. [CrossRef]

21. Zhang, J.; Wang, P.; Zhang, Z.; Xiang, P.; Xia, S. Biosorption Characteristics of Hg(II) from Aqueous Solution by the Biopolymer from Waste Activated Sludge. *Int. J. Environ. Res. Public Health* **2020**, *17*, 1488. [CrossRef]

22. de Freitas, G.R.; da Silva, M.G.C.; Vieira, M.G.A. Biosorption technology for removal of toxic metals: A review of commercial biosorbents and patents. *Environ. Sci. Pollut. Res. Int.* **2019**, *26*, 19097–19118. [CrossRef]

23. Fomina, M.; Gadd, G.M. Biosorption: Current perspectives on concept, definition and application. *Bioresour. Technol.* **2014**, *160*, 3–14. [CrossRef]

24. Blagojev, N.; Kukic, D.; Vasic, V.; Sciban, M.; Prodanovic, J.; Bera, O. A new approach for modelling and optimization of Cu(II) biosorption from aqueous solutions using sugar beet shreds in a fixed-bed column. *J. Hazard. Mater.* **2019**, *363*, 366–375. [CrossRef]

25. Wernke, G.; Fagundes-Klen, M.R.; Vieira, M.F.; Suzaki, P.Y.R.; Souza, H.K.S.; Shimabuku, Q.L.; Bergamasco, R. Mathematical modelling applied to the rate-limiting mass transfer step determination of a herbicide biosorption onto fixed-bed columns. *Environ. Technol.* **2020**, *41*, 638–648. [CrossRef]

26. Aryal, M. Calcium alginate entrapped *Eupatorium adenophorum* Sprengel stems powder for chromium(VI) biosorption in aqueous mediums. *PLoS ONE* **2019**, *14*, e0213477. [CrossRef]

27. Ahmad, A.; Bhat, A.H.; Buang, A. Enhanced biosorption of transition metals by living *Chlorella vulgaris* immobilized in Ca-alginate beads. *Environ. Technol.* **2019**, *40*, 1793–1809. [CrossRef]

28. Paez-Velez, C.; Castro-Mayorga, J.L.; Dussan, J. Effective Gold Biosorption by Electrospun and Electrosprayed Bio-composites with Immobilized *Lysinibacillus sphaericus* CBAM5. *Nanomaterials (Basel)* **2020**, *10*, 408. [CrossRef]

29. Giese, E.C.; Silva, D.D.V.; Costa, A.F.M.; Almeida, S.G.C.; Dussan, K.J. Immobilized microbial nanoparticles for biosorption. *Crit. Rev. Biotechnol.* **2020**, *40*, 653–666. [CrossRef]

30. Kocaman, S. Synthesis and cationic dye biosorption properties of a novel low-cost adsorbent: Coconut waste modified with acrylic and polyacrylic acids. *Int. J. Phytoremediat.* **2020**, *22*, 551–566. [CrossRef]

31. Zhang, J.; Chen, X.; Zhou, J.; Luo, X. Uranium biosorption mechanism model of protonated *Saccharomyces cerevisiae*. *J. Hazard. Mater.* **2020**, *385*, 121588. [CrossRef] [PubMed]

32. Ribeiro, V.R.; Maciel, G.M.; Fachi, M.M.; Pontarolo, R.; Fernandes, I.A.A.; Stafussa, A.P.; Haminiuk, C.W.I. Improvement of phenolic compound bioaccessibility from yerba mate (*Ilex paraguariensis*) extracts after biosorption on *Saccharomyces cerevisiae*. *Food Res. Int.* **2019**, *126*, 108623. [CrossRef] [PubMed]

33. Villen-Guzman, M.; Gutierrez-Pinilla, D.; Gomez-Lahoz, C.; Vereda-Alonso, C.; Rodriguez-Maroto, J.M.; Arhoun, B. Optimization of Ni (II) biosorption from aqueous solution on modified lemon peel. *Environ. Res.* **2019**, *179*, 108849. [CrossRef]

34. Tan, L.; Cui, H.; Xiao, Y.; Xu, H.; Xu, M.; Wu, H.; Dong, H.; Qiu, G.; Liu, X.; Xie, J. Enhancement of platinum biosorption by surface-displaying EC20 on *Escherichia coli*. *Ecotoxicol. Environ. Saf.* **2019**, *169*, 103–111. [CrossRef]

35. Dong, X.B.; Huang, W.; Bian, Y.B.; Feng, X.; Ibrahim, S.A.; Shi, D.F.; Qiao, X.; Liu, Y. Remediation and Mechanisms of Cadmium Biosorption by a Cadmium-Binding Protein from *Lentinula edodes*. *J. Agric. Food Chem.* **2019**, *67*, 11373–11379. [CrossRef]

36. De Oliveira, V.H.; Ullah, I.; Dunwell, J.M.; Tibbett, M. Bioremediation potential of Cd by transgenic yeast expressing a metallothionein gene from *Populus trichocarpa*. *Ecotoxicol. Environ. Saf.* **2020**, *202*, 110917. [CrossRef]

37. Othmani, A.; Kesraoui, A.; HaneneAkrout; Elaissaoui, I.; Seffen, M. Coupling anodic oxidation, biosorption and alternating current as alternative for wastewater purification. *Chemosphere* **2020**, *249*, 126480. [CrossRef]

38. Othmani, A.; Kesraoui, A.; Seffen, M. Removal of phenol from aqueous solution by coupling alternating current with biosorption. *Environ. Sci. Pollut. Res. Int.* **2020**. [CrossRef]

39. de Freitas, G.R.; Vieira, M.G.A.; da Silva, M.G.C. Fixed bed biosorption of silver and investigation of functional groups on acidified biosorbent from algae biomass. *Environ. Sci. Pollut. Res. Int.* **2019**, *26*, 36354–36366. [CrossRef]

40. Debs, K.B.; da Silva, H.D.T.; de Moraes, M.D.L.L.; Carrilho, E.; Lemos, S.G.; Labuto, G. Biosorption of 17alpha-ethinylestradiol by yeast biomass from ethanol industry in the presence of estrone. *Environ. Sci. Pollut. Res. Int.* **2019**, *26*, 28419–28428. [CrossRef]

41. Chen, X.; Zheng, M.; Zhang, G.; Li, F.; Chen, H.; Leng, Y. The nature of dissolved organic matter determines the biosorption capacity of Cu by algae. *Chemosphere* **2020**, *252*, 126465. [CrossRef]

42. do Nascimento Junior, W.J.; da Silva, M.G.C.; Vieira, M.G.A. Competitive biosorption of $Cu^{(2+)}$ and $Ag^{(+)}$ ions on brown macro-algae waste: Kinetic and ion-exchange studies. *Environ. Sci. Pollut. Res. Int.* **2019**, *26*, 23416–23428. [CrossRef]

43. Costa, C.S.D.; Queiroz, B.G.M.; Landers, R.; da Silva, M.G.C.; Vieira, M.G.A. Equilibrium study of binary mixture biosorption of Cr(III) and Zn(II) by dealginated seaweed waste: Investigation of adsorption mechanisms using X-ray photoelectron spectroscopy analysis. *Environ. Sci. Pollut. Res. Int.* **2019**, *26*, 28470–28480. [CrossRef]

44. Rahman, Z.; Thomas, L.; Singh, V.P. Biosorption of heavy metals by a lead (Pb) resistant bacterium, *Staphylococcus hominis* strain AMB-2. *J. Basic Microbiol.* **2019**, *59*, 477–486. [CrossRef]

45. Liu, L.; Liu, J.; Liu, X.; Dai, C.; Zhang, Z.; Song, W.; Chu, Y. Kinetic and equilibrium of U(VI) biosorption onto the resistant bacterium *Bacillus amyloliquefaciens*. *J. Environ. Radioact.* **2019**, *203*, 117–124. [CrossRef]

46. Wang, N.; Qiu, Y.; Xiao, T.; Wang, J.; Chen, Y.; Xu, X.; Kang, Z.; Fan, L.; Yu, H. Comparative studies on Pb(II) biosorption with three spongy microbe-based biosorbents: High performance, selectivity and application. *J. Hazard. Mater.* **2019**, *373*, 39–49. [CrossRef]

47. Gunan Yucel, H.; Aksu, Z.; Yalcinkaya, G.B.; Karatay, S.E.; Donmez, G. A comparative investigation of lithium(I) biosorption properties of *Aspergillus versicolor* and *Kluyveromyces marxianus*. *Water Sci. Technol.* **2020**, *81*, 499–507. [CrossRef]

48. Adekanmbi, E.O.; Giduthuri, A.T.; Carv, B.A.C.; Counts, J.; Moberly, J.G.; Srivastava, S.K. Application of dielectrophoresis towards characterization of rare earth elements biosorption by *Cupriavidus necator*. *Anal. Chim. Acta* **2020**, *1129*, 150–157. [CrossRef]

49. Lagergren, S. About the theory of so-Called adsorption of soluble substance. *Handlingar* **1898**, *24*, 1–39.

50. Blanchard, G.; Maunaye, M.; Martin, G. Removal of heavy metals from waters by means of natural zeolites. *Water Res.* **1984**, *18*, 1501–1507. [CrossRef]

51. Weber, W.J.; Morris, J.C. Kinetics of adsorption on carbon from solutions. *J. Sanit. Eng. Div.* **1963**, *89*, 31–60.

52. Karoui, S.; Ben Arfi, R.; Fernandez-Sanjurjo, M.J.; Nunez-Delgado, A.; Ghorbal, A.; Alvarez-Rodriguez, E. Optimization of synergistic biosorption of oxytetracycline and cadmium from binary mixtures on reed-based beads: Modeling study using Brouers-Sotolongo models. *Environ. Sci. Pollut. Res. Int.* **2020**. [CrossRef]

53. Langmuir, I. The adsorption of gases on plane surfaces of glass, mica and platinum. *J. Am. Chem. Soc.* **1918**, *40*, 1361–1403. [CrossRef]

54. Freundlich, H.M.F. Over the adsorption in solution. *J. Phys. Chem.* **1906**, *57*, 385–470.

55. Giese, E.C. Biosorption as green technology for the recovery and separation of rare earth elements. *World J. Microbiol. Biotechnol.* **2020**, *36*, 52. [CrossRef]

56. Georgin, J.; Franco, D.S.P.; Netto, M.S.; Allasia, D.; Oliveira, M.L.S.; Dotto, G.L. Treatment of water containing methylene by biosorption using Brazilian berry seeds (*Eugenia uniflora*). *Environ. Sci. Pollut. Res. Int.* **2020**, *27*, 20831–20843. [CrossRef]

57. Ribeiro, C.; Scheufele, F.B.; Alves, H.J.; Kroumov, A.D.; Espinoza-Quinones, F.R.; Modenes, A.N.; Borba, C.E. Evaluation of hybrid neutralization/biosorption process for zinc ions removal from automotive battery effluent by dolomite and fish scales. *Environ. Technol.* **2019**, *40*, 2373–2388. [CrossRef]

58. Volesky, B. *Biosorption of Heavy Metals*; Volesky, B., Ed.; CRC Press: Boca Raton, FL, USA, 1990; p. 396.

59. Tsezos, M.; Volesky, B. The mechanism of thorium biosorption by Rhizopus arrhizus. *Biotechnol. Bioeng.* **1982**, *24*, 955–969. [CrossRef]

60. Fisher, R.M.; Gupta, V. *Heavy Metals*; StatPearls Publishing: Treasure Island, FL, USA, 2020.

61. Jaishankar, M.; Tseten, T.; Anbalagan, N.; Mathew, B.B.; Beeregowda, K.N. Toxicity, mechanism and health effects of some heavy metals. *Interdiscip. Toxicol.* **2014**, *7*, 60–72. [CrossRef]

62. Tchounwou, P.B.; Yedjou, C.G.; Patlolla, A.K.; Sutton, D.J. Heavy metal toxicity and the environment. *Exp. Suppl.* **2012**, *101*, 133–164. [CrossRef]

63. Qin, H.; Hu, T.; Zhai, Y.; Lu, N.; Aliyeva, J. The improved methods of heavy metals removal by biosorbents: A review. *Environ. Pollut.* **2020**, *258*, 113777. [CrossRef] [PubMed]

64. Dinh, V.P.; Xuan, T.D.; Hung, N.Q.; Luu, T.T.; Do, T.T.; Nguyen, T.D.; Nguyen, V.D.; Anh, T.T.K.; Tran, N.Q. Primary biosorption mechanism of lead (II) and cadmium (II) cations from aqueous solution by pomelo (*Citrus maxima*) fruit peels. *Environ. Sci. Pollut. Res. Int.* **2020**. [CrossRef] [PubMed]

65. Sedlakova-Kadukova, J.; Kopcakova, A.; Gresakova, L.; Godany, A.; Pristas, P. Bioaccumulation and biosorption of zinc by a novel *Streptomyces* K11 strain isolated from highly alkaline aluminium brown mud disposal site. *Ecotoxicol. Environ. Saf.* **2019**, *167*, 204–211. [CrossRef] [PubMed]

66. Moreira, V.R.; Lebron, Y.A.R.; Lange, L.C.; Santos, L.V.S. Simultaneous biosorption of Cd(II), Ni(II) and Pb(II) onto a brown macroalgae *Fucus vesiculosus*: Mono- and multi-component isotherms, kinetics and thermodynamics. *J. Environ. Manag.* **2019**, *251*, 109587. [CrossRef]

67. Libatique, M.J.H.; Lee, M.C.; Yeh, H.Y.; Jhang, F.J. Total and inorganic arsenic biosorption by *Sarcodia suiae* (*Rhodophyta*), as affected by controlled environmental conditions. *Chemosphere* **2020**, *248*, 126084. [CrossRef]

68. Hu, X.; Cao, J.; Yang, H.; Li, D.; Qiao, Y.; Zhao, J.; Zhang, Z.; Huang, L. Pb^{2+} biosorption from aqueous solutions by live and dead biosorbents of the hydrocarbon-degrading strain *Rhodococcus* sp. HX-2. *PLoS ONE* **2020**, *15*, e0226557. [CrossRef]

69. Dai, Q.H.; Bian, X.Y.; Li, R.; Jiang, C.B.; Ge, J.M.; Li, B.L.; Ou, J. Biosorption of lead(II) from aqueous solution by lactic acid bacteria. *Water Sci. Technol.* **2019**, *79*, 627–634. [CrossRef]

70. Barquilha, C.E.R.; Cossich, E.S.; Tavares, C.R.G.; da Silva, E.A. Biosorption of nickel(II) and copper(II) ions from synthetic and real effluents by alginate-based biosorbent produced from seaweed *Sargassum* sp. *Environ. Sci. Pollut. Res. Int.* **2019**, *26*, 11100–11112. [CrossRef]

71. Lin, Z.; Li, J.; Luan, Y.; Dai, W. Application of algae for heavy metal adsorption: A 20-year meta-analysis. *Ecotoxicol. Environ. Saf.* **2020**, *190*, 110089. [CrossRef]

72. Ojima, Y.; Kosako, S.; Kihara, M.; Miyoshi, N.; Igarashi, K.; Azuma, M. Recovering metals from aqueous solutions by biosorption onto phosphorylated dry baker's yeast. *Sci. Rep.* **2019**, *9*, 225. [CrossRef]

73. Zoroufchi, B.K.; Motalebi, D.A.; McPhedran, K.N.; Soltan, J. Treatment of aqueous arsenic—A review of biosorbent preparation methods. *J. Environ. Manag.* **2020**, *273*, 111126. [CrossRef] [PubMed]

74. Zhang, C.; Ren, H.X.; Zhong, C.Q.; Wu, D. Biosorption of Cr(VI) by immobilized waste biomass from polyglutamic acid production. *Sci. Rep.* **2020**, *10*, 3705. [CrossRef] [PubMed]

75. Rambabu, K.; Bharath, G.; Banat, F.; Show, P.L. Biosorption performance of date palm empty fruit bunch wastes for toxic hexavalent chromium removal. *Environ. Res.* **2020**, *187*, 109694. [CrossRef] [PubMed]

76. Kumar, S.; Shahnaz, T.; Selvaraju, N.; Rajaraman, P.V. Kinetic and thermodynamic studies on biosorption of Cr(VI) on raw and chemically modified *Datura stramonium* fruit. *Environ. Monit. Assess.* **2020**, *192*, 248. [CrossRef] [PubMed]

77. Kalola, V.; Desai, C. Biosorption of Cr(VI) by *Halomonas* sp. DK4, a halotolerant bacterium isolated from chrome electroplating sludge. *Environ. Sci. Pollut. Res. Int.* **2020**, *27*, 27330–27344. [CrossRef] [PubMed]

78. Mishra, A.; Gupta, B.; Kumar, N.; Singh, R.; Varma, A.; Thakur, I.S. Synthesis of calcite-based bio-composite biochar for enhanced biosorption and detoxification of chromium Cr (VI) by *Zhihengliuella* sp. ISTPL4. *Bioresour. Technol.* **2020**, *307*, 123262. [CrossRef]

79. Aranda-Garcia, E.; Cristiani-Urbina, E. Hexavalent chromium removal and total chromium biosorption from aqueous solution by *Quercus crassipes* acorn shell in a continuous up-flow fixed-bed column: Influencing parameters, kinetics, and mechanism. *PLoS ONE* **2020**, *15*, e0227953. [CrossRef]

80. Antony, G.S.; Manna, A.; Baskaran, S.; Puhazhendi, P.; Ramchary, A.; Niraikulam, A.; Ramudu, K.N. Non-enzymatic reduction of Cr (VI) and it's effective biosorption using heat-inactivated biomass: A fermentation waste material. *J. Hazard. Mater.* **2020**, *392*, 122257. [CrossRef]

81. Wang, J.; Xie, Q.; Li, A.; Liu, X.; Yu, F.; Ji, J. Biosorption of hexavalent chromium from aqueous solution by polyethyleneimine-modified ultrasonic-assisted acid hydrochar from *Sargassum horneri*. *Water Sci. Technol.* **2020**, *81*, 1114–1129. [CrossRef]

82. Prabhu, S.G.; Srinikethan, G.; Hegde, S. Spontaneous Cr(VI) and Cd(II) biosorption potential of native pinnae tissue of *Pteris vittata* L., a tropical invasive pteridophyte. *Int. J. Phytoremediat.* **2019**, *21*, 380–390. [CrossRef]

83. Gocenoglu, A. Kinetic and thermodynamic studies of the biosorption of Cr (VI) in aqueous solutions by *Agaricus campestris*. *Environ. Technol.* **2019**, 1–9. [CrossRef]

84. da Rocha Ferreira, G.L.; Vendruscolo, F.; Antoniosi Filho, N.R. Biosorption of hexavalent chromium by *Pleurotus ostreatus*. *Heliyon* **2019**, *5*, e01450. [CrossRef]

85. Chang, J.; Deng, S.; Liang, Y.; Chen, J. Cr(VI) removal performance from aqueous solution by *Pseudomonas* sp. strain DC-B3 isolated from mine soil: Characterization of both Cr(VI) bioreduction and total Cr biosorption processes. *Environ. Sci. Pollut. Res. Int.* **2019**, *26*, 28135–28145. [CrossRef]

86. Campana-Perez, J.F.; Portero Barahona, P.; Martin-Ramos, P.; Carvajal Barriga, E.J. Ecuadorian yeast species as microbial particles for Cr(VI) biosorption. *Environ. Sci. Pollut. Res. Int.* **2019**, *26*, 28162–28172. [CrossRef]

87. Tyagi, B.; Gupta, B.; Thakur, I.S. Biosorption of Cr (VI) from aqueous solution by extracellular polymeric substances (EPS) produced by *Parapedobacter* sp. ISTM3 strain isolated from Mawsmai cave, Meghalaya, India. *Environ. Res.* **2020**, *191*, 110064. [CrossRef]

88. Devi, A.P.; Mishra, P.M. Biosorption of dysprosium (III) using raw and surface-modified bark powder of *Mangifera indica*: Isotherm, kinetic and thermodynamic studies. *Environ. Sci. Pollut. Res. Int.* **2019**, *26*, 6545–6556. [CrossRef]

89. Wang, L.; Li, Z.; Wang, Y.; Brookes, P.C.; Wang, F.; Zhang, Q.; Xu, J.; Liu, X. Performance and mechanisms for remediation of Cd(II) and As(III) co-contamination by magnetic biochar-microbe biochemical composite: Competition and synergy effects. *Sci. Total Environ.* **2020**, *750*, 141672. [CrossRef]

90. Huang, H.; Jia, Q.; Jing, W.; Dahms, H.U.; Wang, L. Screening strains for microbial biosorption technology of cadmium. *Chemosphere* **2020**, *251*, 126428. [CrossRef]

91. Wang, Q.; Li, Q.; Lin, Y.; Hou, Y.; Deng, Z.; Liu, W.; Wang, H.; Xia, Z. Biochemical and genetic basis of cadmium biosorption by *Enterobacter ludwigii* LY6, isolated from industrial contaminated soil. *Environ. Pollut.* **2020**, *264*, 114637. [CrossRef]

92. Boczonadi, I.; Torok, Z.; Jakab, A.; Konya, G.; Gyurcso, K.; Baranyai, E.; Szoboszlai, Z.; Donczo, B.; Fabian, I.; Leiter, E.; et al. Increased Cd$(^{2+})$ biosorption capability of *Aspergillus nidulans* elicited by crpA deletion. *J. Basic Microbiol.* **2020**, *60*, 574–584. [CrossRef]

93. Ibuot, A.; Webster, R.E.; Williams, L.E.; Pittman, J.K. Increased metal tolerance and bioaccumulation of zinc and cadmium in *Chlamydomonas reinhardtii* expressing a AtHMA4 C-terminal domain protein. *Biotechnol. Bioeng.* **2020**. [CrossRef]

94. Jin, Z.; Xie, L.; Zhang, T.; Liu, L.; Black, T.; Jones, K.C.; Zhang, H.; Wang, X.; Jin, N.; Zhang, D. Interrogating cadmium and lead biosorption mechanisms by *Simplicillium chinense* via infrared spectroscopy. *Environ. Pollut.* **2020**, *263*, 114419. [CrossRef]

95. Yuan, W.; Cheng, J.; Huang, H.; Xiong, S.; Gao, J.; Zhang, J.; Feng, S. Optimization of cadmium biosorption by *Shewanella putrefaciens* using a Box-Behnken design. *Ecotoxicol. Environ. Saf.* **2019**, *175*, 138–147. [CrossRef]

96. Sun, H.; Wang, X.; Wang, R.; Zhang, Y.; Wang, X. Biosorption of Cd$(^{2+})$ from aqueous solution by Ca$(^{2+})$/Mg$(^{2+})$ type *Citrus paradisi* Macf. peel biosorbents. *Water Sci. Technol.* **2019**, *80*, 1205–1212. [CrossRef] [PubMed]

97. Le, B.; Yang, S.H. Biosorption of cadmium by potential probiotic *Pediococcus pentosaceus* using in vitro digestion model. *Biotechnol. Appl. Biochem.* **2019**, *66*, 673–680. [CrossRef]

98. Lu, N.; Hu, T.; Zhai, Y.; Qin, H.; Aliyeva, J.; Zhang, H. Fungal cell with artificial metal container for heavy metals biosorption: Equilibrium, kinetics study and mechanisms analysis. *Environ. Res.* **2020**, *182*, 109061. [CrossRef]

99. Mohapatra, R.K.; Parhi, P.K.; Pandey, S.; Bindhani, B.K.; Thatoi, H.; Panda, C.R. Active and passive biosorption of Pb(II) using live and dead biomass of marine bacterium *Bacillus xiamenensis* PbRPSD202: Kinetics and isotherm studies. *J. Environ. Manag.* **2019**, *247*, 121–134. [CrossRef]

100. Qiao, W.; Zhang, Y.; Xia, H.; Luo, Y.; Liu, S.; Wang, S.; Wang, W. Bioimmobilization of lead by Bacillus subtilis X3 biomass isolated from lead mine soil under promotion of multiple adsorption mechanisms. *R. Soc. Open Sci.* **2019**, *6*, 181701. [CrossRef]

101. Jin, Z.; Deng, S.; Wen, Y.; Jin, Y.; Pan, L.; Zhang, Y.; Black, T.; Jones, K.C.; Zhang, H.; Zhang, D. Application of *Simplicillium chinense* for Cd and Pb biosorption and enhancing heavy metal phytoremediation of soils. *Sci. Total Environ.* **2019**, *697*, 134148. [CrossRef]

102. Imran, M.; Anwar, K.; Akram, M.; Shah, G.M.; Ahmad, I.; Samad Shah, N.; Khan, Z.U.H.; Rashid, M.I.; Akhtar, M.N.; Ahmad, S.; et al. Biosorption of Pb(II) from contaminated water onto *Moringa oleifera* biomass: Kinetics and equilibrium studies. *Int. J. Phytoremediat.* **2019**, *21*, 777–789. [CrossRef]

103. Akram, M.; Khan, B.; Imran, M.; Ahmad, I.; Ajaz, H.; Tahir, M.; Rabbani, F.; Kaleem, I.; Nadeem Akhtar, M.; Ahmad, N.; et al. Biosorption of lead by cotton shells powder: Characterization and equilibrium modeling study. *Int. J. Phytoremediat.* **2019**, *21*, 138–144. [CrossRef] [PubMed]

104. Kumar, M.; Singh, A.K.; Sikandar, M. Biosorption of Hg (II) from aqueous solution using algal biomass: Kinetics and isotherm studies. *Heliyon* **2020**, *6*, e03321. [CrossRef]

105. Fathollahi, A.; Coupe, S.J.; El-Sheikh, A.H.; Sanudo-Fontaneda, L.A. The biosorption of mercury by permeable pavement biofilms in stormwater attenuation. *Sci. Total Environ.* **2020**, *741*, 140411. [CrossRef]

106. Chen, C.; Hu, J.; Wang, J. Biosorption of uranium by immobilized *Saccharomyces cerevisiae*. *J. Environ. Radioact.* **2020**, *213*, 106158. [CrossRef]

107. Vieira, L.C.; de Araujo, L.G.; de Padua Ferreira, R.V.; da Silva, E.A.; Canevesi, R.L.S.; Marumo, J.T. Uranium biosorption by *Lemna* sp. and *Pistia stratiotes*. *J. Environ. Radioact.* **2019**, *203*, 179–186. [CrossRef]

108. Han, J.; Hu, L.; He, L.; Ji, K.; Liu, Y.; Chen, C.; Luo, X.; Tan, N. Preparation and uranium (VI) biosorption for tri-amidoxime modified marine fungus material. *Environ. Sci. Pollut. Res. Int.* **2020**. [CrossRef]

109. Dai, Y.; Zhou, L.; Tang, X.; Xi, J.; Ouyang, J.; Liu, Z.; Huang, G.; Adesina, A.A. Macroporous ion-imprinted chitosan foams for the selective biosorption of U(VI) from aqueous solution. *Int. J. Biol. Macromol.* **2020**, *164*, 4155–4164. [CrossRef]

110. Ferreira, R.V.P.; de Araujo, L.G.; Canevesi, R.L.S.; da Silva, E.A.; Ferreira, E.G.A.; Palmieri, M.C.; Marumo, J.T. The use of rice and coffee husks for biosorption of U (total), (241)Am, and (137)Cs in radioactive liquid organic waste. *Environ. Sci. Pollut. Res. Int.* **2020**, *27*, 36651–36663. [CrossRef]

111. Ramasamy, K.P.; Rajasabapathy, R.; Lips, I.; Mohandass, C.; James, R.A. Genomic features and copper biosorption potential of a new *Alcanivorax* sp. VBW004 isolated from the shallow hydrothermal vent (Azores, Portugal). *Genomics* **2020**, *112*, 3268–3273. [CrossRef]

112. Ghoniem, A.A.; El-Naggar, N.E.; Saber, W.I.A.; El-Hersh, M.S.; El-Khateeb, A.Y. Statistical modeling-approach for optimization of Cu(2+) biosorption by *Azotobacter nigricans* NEWG-1; characterization and application of immobilized cells for metal removal. *Sci. Rep.* **2020**, *10*, 9491. [CrossRef]

113. do Nascimento, J.M.; de Oliveira, J.D.; Rizzo, A.C.L.; Leite, S.G.F. Biosorption Cu (II) by the yeast *Saccharomyces cerevisiae*. *Biotechnol. Rep. (Amst)* **2019**, *21*, e00315. [CrossRef] [PubMed]

114. Lacerda, E.C.M.; Dos Passos Galluzzi Baltazar, M.; Dos Reis, T.A.; do Nascimento, C.A.O.; Correa, B.; Gimenes, L.J. Copper biosorption from an aqueous solution by the dead biomass of *Penicillium ochrochloron*. *Environ. Monit. Assess.* **2019**, *191*, 247. [CrossRef] [PubMed]

115. Ali, M.M.; Bhakta, J.N. Biosorption of zinc from aqueous solution using leaves of *Corchorus olitorius* as a low-cost biosorbent. *Water Environ. Res.* **2020**, *92*, 821–828. [CrossRef] [PubMed]

116. Liu, X.; Han, B.; Su, C.L.; Han, Q.; Chen, K.J.; Chen, Z.Q. Optimization and mechanisms of biosorption process of Zn(II) on rape straw powders in aqueous solution. *Environ. Sci. Pollut. Res. Int.* **2019**, *26*, 32151–32164. [CrossRef]

117. Tariq, A.; Ullah, U.; Asif, M.; Sadiq, I. Biosorption of arsenic through bacteria isolated from Pakistan. *Int. Microbiol.* **2019**, *22*, 59–68. [CrossRef]

118. Li, D.; Li, R.; Ding, Z.; Ruan, X.; Luo, J.; Chen, J.; Zheng, J.; Tang, J. Discovery of a novel native bacterium of *Providencia* sp. with high biosorption and oxidation ability of manganese for bioleaching of heavy metal contaminated soils. *Chemosphere* **2020**, *241*, 125039. [CrossRef]

119. Noormohamadi, H.R.; Fat'hi, M.R.; Ghaedi, M.; Ghezelbash, G.R. Potentiality of white-rot fungi in biosorption of nickel and cadmium: Modeling optimization and kinetics study. *Chemosphere* **2019**, *216*, 124–130. [CrossRef]

120. Li, H.; Dong, W.; Liu, Y.; Zhang, H.; Wang, G. Enhanced Biosorption of Nickel Ions on Immobilized Surface-Engineered Yeast Using Nickel-Binding Peptides. *Front. Microbiol.* **2019**, *10*, 1254. [CrossRef]

121. Lee, K.Y.; Lee, S.H.; Lee, J.E.; Lee, S.Y. Biosorption of radioactive cesium from contaminated water by microalgae *Haematococcus pluvialis* and *Chlorella vulgaris*. *J. Environ. Manag.* **2019**, *233*, 83–88. [CrossRef]

122. Giese, E.C.; Dekker, R.F.H.; Barbosa-Dekker, A.M. Biosorption of lanthanum and samarium by viable and autoclaved mycelium of *Botryosphaeria rhodina* MAMB-05. *Biotechnol. Prog.* **2019**, *35*, e2783. [CrossRef]

123. Kalak, T.; Dudczak-Halabuda, J.; Tachibana, Y.; Cierpiszewski, R. Effective use of elderberry (*Sambucus nigra*) pomace in biosorption processes of Fe(III) ions. *Chemosphere* **2020**, *246*, 125744. [CrossRef] [PubMed]

124. Krishna Kanamarlapudi, S.L.R.; Muddada, S. Structural Changes of *Bacillus subtilis* Biomass on Biosorption of Iron (II) from Aqueous Solutions: Isotherm and Kinetic Studies. *Pol. J. Microbiol.* **2019**, *68*, 549–558. [CrossRef] [PubMed]

125. Lapworth, D.J.; Baran, N.; Stuart, M.E.; Ward, R.S. Emerging organic contaminants in groundwater: A review of sources, fate and occurrence. *Environ. Pollut.* **2012**, *163*, 287–303. [CrossRef] [PubMed]

126. Garcia, J.; Garcia-Galan, M.J.; Day, J.W.; Boopathy, R.; White, J.R.; Wallace, S.; Hunter, R.G. A review of emerging organic contaminants (EOCs), antibiotic resistant bacteria (ARB), and antibiotic resistance genes (ARGs) in the environment: Increasing removal with wetlands and reducing environmental impacts. *Bioresour. Technol.* **2020**, *307*, 123228. [CrossRef] [PubMed]

127. Sun, W.; Sun, W.; Wang, Y. Biosorption of Direct Fast Scarlet 4BS from aqueous solution using the green-tide-causing marine algae *Enteromorpha prolifera*. *Spectrochim. Acta Part A Mol. Biomol. Spectrosc.* **2019**, *223*, 117347. [CrossRef]

128. Dada, A.O.; Adekola, F.A.; Odebunmi, E.O.; Dada, F.E.; Bello, O.M.; Akinyemi, B.A.; Bello, O.S.; Umukoro, O.G. Sustainable and low-cost *Ocimum gratissimum* for biosorption of indigo carmine dye: Kinetics, isotherm, and thermodynamic studies. *Int. J. Phytoremediat.* **2020**, 1–14. [CrossRef]

129. Sunsandee, N.; Ramakul, P.; Phatanasri, S.; Pancharoen, U. Biosorption of dicloxacillin from pharmaceutical waste water using tannin from Indian almond leaf: Kinetic and equilibrium studies. *Biotechnol. Rep. (Amst)* **2020**, *27*, e00488. [CrossRef]

130. Pi, S.; Li, A.; Cui, D.; Su, Z.; Feng, L.; Ma, F.; Yang, J. Biosorption behavior and mechanism of sulfonamide antibiotics in aqueous solution on extracellular polymeric substances extracted from *Klebsiella* sp. J1. *Bioresour. Technol.* **2019**, *272*, 346–350. [CrossRef]

131. Ezekoye, O.M.; Akpomie, K.G.; Eze, S.I.; Chukwujindu, C.N.; Ani, J.U.; Ujam, O.T. Biosorptive interaction of alkaline modified *Dialium guineense* seed powders with ciprofloxacin in contaminated solution: Central composite, kinetics, isotherm, thermodynamics, and desorption. *Int. J. Phytoremediat.* **2020**, *22*, 1028–1037. [CrossRef]

132. Parlayici, S.; Pehlivan, E. Biosorption of methylene blue and malachite green on biodegradable magnetic *Cortaderia selloana* flower spikes: Modeling and equilibrium study. *Int. J. Phytoremediat.* **2020**, 1–15. [CrossRef]

133. de Araujo, T.P.; Tavares, F.O.; Vareschini, D.T.; Barros, M. Biosorption mechanisms of cationic and anionic dyes in a low-cost residue from brewer's spent grain. *Environ. Technol.* **2020**, 1–16. [CrossRef] [PubMed]

134. Silva, F.; Nascimento, L.; Brito, M.; da Silva, K.; Paschoal, W., Jr.; Fujiyama, R. Biosorption of Methylene Blue Dye Using Natural Biosorbents Made from Weeds. *Materials* **2019**, *12*, 2486. [CrossRef] [PubMed]

135. Lebron, Y.A.R.; Moreira, V.R.; de Souza Santos, L.V. Biosorption of methylene blue and eriochrome black T onto the brown macroalgae *Fucus vesiculosus*: Equilibrium, kinetics, thermodynamics and optimization. *Environ. Technol.* **2019**, 1–19. [CrossRef]

136. Bouzikri, S.; Ouasfi, N.; Benzidia, N.; Salhi, A.; Bakkas, S.; Khamliche, L. Marine alga *"Bifurcaria bifurcata"*: Biosorption of Reactive Blue 19 and methylene blue from aqueous solutions. *Environ. Sci. Pollut. Res. Int.* **2020**, *27*, 33636–33648. [CrossRef] [PubMed]

137. Grassi, P.; Reis, C.; Drumm, F.C.; Georgin, J.; Tonato, D.; Escudero, L.B.; Kuhn, R.; Jahn, S.L.; Dotto, G.L. Biosorption of crystal violet dye using inactive biomass of the fungus *Diaporthe schini*. *Water Sci. Technol.* **2019**, *79*, 709–717. [CrossRef]

138. Azari, A.; Noorisepehr, M.; Dehghanifard, E.; Karimyan, K.; Hashemi, S.Y.; Kalhori, E.M.; Norouzi, R.; Agarwal, S.; Gupta, V.K. Experimental design, modeling and mechanism of cationic dyes biosorption on to magnetic chitosan-lutaraldehyde composite. *Int. J. Biol. Macromol.* **2019**, *131*, 633–645. [CrossRef]

139. Deniz, F.; Yildiz, H. Taguchi DoE methodology for modeling of synthetic dye biosorption from aqueous effluents: Parametric and phenomenological studies. *Int. J. Phytoremediat.* **2019**, *21*, 1065–1071. [CrossRef]

140. Selvakumar, A.; Rangabhashiyam, S. Biosorption of Rhodamine B onto novel biosorbents from *Kappaphycus alvarezii*, *Gracilaria salicornia* and *Gracilaria edulis*. *Environ. Pollut.* **2019**, *255*, 113291. [CrossRef] [PubMed]

141. Liu, F.; Zhang, X.; Wang, M.; Guo, L.; Yang, Y.; Zhao, M. Biosorption of sterols from tobacco waste extract using living and dead of newly isolated fungus *Aspergillus fumigatus* strain LSD-1. *Biosci. Biotechnol. Biochem.* **2020**, *84*, 1521–1528. [CrossRef] [PubMed]

142. Coelho, C.M.; de Andrade, J.R.; da Silva, M.G.C.; Vieira, M.G.A. Removal of propranolol hydrochloride by batch biosorption using remaining biomass of alginate extraction from *Sargassum filipendula* algae. *Environ. Sci. Pollut. Res. Int.* **2020**, *27*, 16599–16611. [CrossRef] [PubMed]

143. Bo, L.G.; Almeida, R.M.; Cardoso, C.M.M.; Zavarize, D.G.; Brum, S.S.; Mendonca, A.R.V. Acetylsalicylic acid biosorption onto fungal-bacterial biofilm supported on activated carbons: An investigation via batch and fixed-bed experiments. *Environ. Sci. Pollut. Res. Int.* **2019**, *26*, 28962–28976. [CrossRef] [PubMed]

144. Silva, A.; Coimbra, R.N.; Escapa, C.; Figueiredo, S.A.; Freitas, O.M.; Otero, M. Green Microalgae *Scenedesmus Obliquus* Utilization for the Adsorptive Removal of Nonsteroidal Anti-Inflammatory Drugs (NSAIDs) from Water Samples. *Int. J. Environ. Res. Public Health* **2020**, *17*, 3707. [CrossRef] [PubMed]

145. De Gisi, S.; Lofrano, G.; Grassi, M.; Notarnicola, M. Characteristics and adsorption capacities of low-cost sorbents for wastewater treatment: A review. *Sustain. Mater. Technol.* **2016**, *9*, 10–40. [CrossRef]

146. Crini, G.; Lichtfouse, E.; Wilson, L.D.; Morin-Crini, N. Conventional and non-conventional adsorbents for wastewater treatment. *Environ. Chem. Lett.* **2019**, *17*, 195–213. [CrossRef]

147. Zhu, C.; Zhai, X.; Xi, Y.; Wang, J.; Kong, F.; Zhao, Y.; Chi, Z. Progress on the development of floating photobioreactor for microalgae cultivation and its application potential. *World J. Microbiol. Biotechnol.* **2019**, *35*, 190. [CrossRef]

Publisher's Note: MDPI stays neutral with regard to jurisdictional claims in published maps and institutional affiliations.

 *processes*

Article

Batch and Fixed-Bed Biosorption of Pb (II) Using Free and Alginate-Immobilized *Spirulina*

Maria Villen-Guzman [1,*], Carlos Jiménez [2] and Jose Miguel Rodriguez-Maroto [1]

[1] Department of Chemical Engineering, Faculty of Sciences, University of Malaga, 29071 Malaga, Spain; maroto@uma.es
[2] Department of Ecology, Faculty of Sciences, University of Malaga, 29071 Malaga, Spain; carlosj@uma.es
* Correspondence: mvillen@uma.es

Abstract: The valorization of *Spirulina* as a potential biosorption material to treat contaminated wastewater was evaluated. Batch experiments were conducted to study the influence of pH value and ionic strength on the biosorption capacity of *Spirulina*. Higher removal capacity was observed at pH 5.2, while higher ionic strength was found to result in lower adsorption capacity, which suggests that ion exchange is a relevant mechanism for Pb (II) adsorption on *Spirulina*. The immobilization of *Spirulina* on alginate beads was found not only to increase the adsorption capacity, but also to overcome limitations such as unacceptable pressure drops on column systems. The Langmuir model was the most appropriate model to describe the biosorption equilibrium of lead by free and immobilized *Spirulina*. The experimental breakthrough curves were evaluated using the Thomas, Bohart-Adams, and dose-response models. The experimental results were most properly described by the dose-response model, which is consistent with previous results. The adsorption capacity of *Spirulina* was found to increase linearly with the influent lead concentration (in the range 4–20 mg L^{-1}) at 1.6 mL min^{-1} flow rate. Batch and column experiments were compared to better understand the biosorption process. The promising results obtained indicate the potential use of *Spirulina* immobilized on alginate beads to treat industrial wastewater polluted with toxic metals.

Keywords: biosorption; *Spirulina*; alginate; immobilization; fixed-bed column

Citation: Villen-Guzman, M.; Jiménez, C.; Rodriguez-Maroto, J.M. Batch and Fixed-Bed Biosorption of Pb (II) Using Free and Alginate-Immobilized *Spirulina*. *Processes* **2021**, *9*, 466. https://doi.org/10.3390/pr9030466

Academic Editor: Jose Enrique Torres Vaamonde

Received: 14 February 2021
Accepted: 1 March 2021
Published: 5 March 2021

Publisher's Note: MDPI stays neutral with regard to jurisdictional claims in published maps and institutional affiliations.

1. Introduction

The increasing pollution of the environment by heavy metals due to industrial activity poses serious risks for human health and living organisms. Lead, which is non-biodegradable and toxic even at low concentrations, is one of the most serious environmental pollutants [1]. The presence of lead in various types of wastewater should be controlled according to the standard for the permitted amount of metal determined by the Environmental Protection Agency [2].

Several conventional technologies have been proposed to remove lead from aquatic environments. Recently, biosorption has been proposed as an emerging and low-cost alternative based on the sorption of dissolved pollutants on a biomaterial. This technology overcomes the most relevant drawbacks of conventional methods, mainly the indirect disposal of toxic metal sludge and the limited adsorption efficiency at low metal concentrations [3]. Biomaterials, such as bacteria, algae, fungi, and agricultural wastes, have been proposed as low-cost biosorbents for the removal of heavy metals from wastewater [4]. Among these biomaterials, algae have been widely suggested as ideal biosorbents for the removal of toxic metals from water effluents [5]. The presence of negatively charged groups on its surface, such as amino, hydroxyl, carboxyl, sulfhydryl, and sulfonate, allows the binding of heavy metals on algal biomass [6].

Brouers and Al-Musawi proposed the use of a fresh mixture of green and blue-green algae as a biosorbent for lead removal contained in aqueous solutions. They concluded that algal biomass is a promising biosorbent with a maximum removal efficiency of 98%

at 40 °C and pH 3 [7]. *Ecklonia radiata*, a brown marine algae, was found to have a much higher value of adsorption capacity for lead than other conventional adsorbents such as powder-activated carbon and natural zeolites [8]. A green marine alga, *Ulva lactuca*, has also been proposed as an effective and natural biosorbent for heavy metal removal under acidic pH conditions [9]. Verma et al. evaluated the role of the brown marine alga *Sargassum filipendula* in the removal of lead from industrial wastewater. They associated the efficiency of the brown algae with its high content in acidic polysaccharide, usually Ca and Na alginates [10].

Spirulina (a blue-green algae) has been also proposed as an ideal biosorbent not only due to its fast growth, but also because it contains a wide range of functional groups, i.e., carboxyl, hydroxyl, sulfate, and other charged groups [11–17]. Şeker et al. evaluated the competitive biosorption of Pb, Cd and Ni, concluding that *Spirulina* has a higher selectivity toward Pb (II) ions. They suggested the application of *Spirulina* as a biosorbent to be used in large-scale batch biosorption systems [18]. Several studies focused on biosorbents have concluded their higher affinity for Pb biosorption in multi-metal systems associated with its ionic properties, i.e., electronegativity, ionic radius, and redox potential [19]. In recent studies, the chemical modification of *Spirulina* using sulfuric acid was proven to allow a strong bond between the metal ions and the modified biosorbent [20,21]. The chemical treatments, such as surface modification through exposing the biosorbent to acid solutions, were focused on improving metal biosorption efficiency and removing soluble organic compounds [22].

The usage of algae as a biosorbent presents some limitations in separating the biomass from the effluent, which produces unacceptable pressure drops. These problems are associated with the physical characteristics of the biomass (i.e., small particle size with low density and poor rigidity) [23]. The use of immobilization or cross-linking technologies could overcome these limitations by providing mechanical strength, rigidity, ideal size, and porous characteristics. The selection of a suitable carrier material is crucial to an effective immobilization. Due to their low cost, simplicity, and high biosorption capacity, alginate-based systems have been widely proposed as immobilization matrixes [24–27].

In this study, *Spirulina* was tested as a biosorbent to remove lead from an aqueous solution. The influence of relevant operational parameters, such as initial pH value of the aqueous solution and ionic strength, was evaluated at batch system. Adsorption isotherm models were applied to experimental data obtained from the biosorption of Pb (II) on free and alginate-immobilized biomass. The assessment of the biosorption process of Pb (II) on *Spirulina* immobilized in calcium alginate beads was performed in a laboratory scale fixed-bed column. The breakthrough curve was used to study the efficiency of the continuous process with the proposed biosorbent.

2. Materials and Methods

2.1. Cultivation of Spirulina sp.

A biosorbent for the removal of lead from aqueous solutions was prepared from the microalga Arthrospira (*Spirulina*) platensis, which was obtained from the Spanish Bank of Algae (BEA 0007B), University of Las Palmas de Gran Canaria. The isolation and cultivation of *Spirulina* was carried out at laboratory scale using Zarrouk medium [28]. The fresh biomass obtained, after filtration through 20 µm mesh, was washed with deionized water and frozen at −80 °C before lyophilization.

Lead solutions were obtained using $Pb(NO_3)_2$ in deionized water. The pH value and the ionic strength were controlled by adding diluted solutions of 0.1 M HNO_3 and 0.01 and 0.1 M $NaNO_3$, respectively.

2.2. Immobilization of Spirulina in Alginate Gel

The immobilization of *Spirulina* was carried out following the method described by Lu and Wilkins [29]. A *Spirulina* biomass (1 g) and deionized water (25 mL) were mixed with the same mass of sodium alginate (from *Macrocystis pyrifera*) (Sigma-Aldrich, St. Louis,

MO, USA) dissolved in deionized water (25 mL). The resulting solution was heated to 85 °C for 15 min. With the aim of obtaining beads with diameters ranging between 1.5 and 2.0 mm, the mixture was dropped through a syringe with an internal diameter of 2 mm into a solution of 0.5 M $CaCl_2$. After 24 h, the solution was rinsed with deionized water and immersed in a solution of 0.5 M HCl for more than 24 h. Finally, the beads were washed again with deionized water.

2.3. Batch Biosorption Studies

Batch experiments were proposed to evaluate the adsorption of Pb (II) on free and immobilized *Spirulina*. The experiments with free biosorbent were conducted by adding a known mass of *Spirulina* to 1 L of aqueous $Pb(NO_3)_2$ solutions of different concentrations. Following the guidelines published by the Organization for Economic Cooperation and Development (OECD) [30], the solid to liquid ratio (S/L) was 50 mg L^{-1}. The samples were stirred in an orbital shaker at 120 rpm at 25 °C. After completion, the mixture was filtered using Whatman GF/C filters (pore size 0.6 μm, Whatman, Maidstone, UK) through vacuum filtration. Each experiment was performed in triplicate using the same operational conditions. The effect of the initial pH of the solutions, i.e., 2, 3, 4, 4.5, and 5.23 (the pH value of the original Pb solution, not adjusted), was evaluated using Pb (II) solutions with initial concentrations in the range 0.6 to 5.6 mg L^{-1}. For isotherm studies, the initial Pb (II) concentration was varied from 0.6 to 5.6 mg L^{-1} at pH 5.23 for 72 h. The effect of ionic strength on the biosorption of Pb (II) by *Spirulina* was evaluated by adding 0.01 and 0.1 M $NaNO_3$. The concentration of Pb (II) was varied from 0.6 to 5.6 mg L^{-1} to obtain isotherms with different ionic strengths.

The behavior of *Spirulina* immobilized on alginate gel was also evaluated by performing batch experiments. The essays were conducted by immersing 35 mg of alginate or *Spirulina* immobilized on alginate gel in 100 mL of aqueous $Pb(NO_3)_2$ solutions of different concentrations (from 20 to 150 mg L^{-1}) until the equilibrium was reached. The final concentration of Pb (II) in the aqueous solution was analyzed by atomic absorption spectroscopy (AAS) (Varian SpectrAA-110, Varian, Palo Alto, CA, USA).

With the aim of determining the total content of Pb (II) in solid samples, microwave-assisted acid digestion was conducted. The concentration of Pb (II) was also determined by AAS (Varian SpectrAA-110). Each experiment was performed in triplicate under the same experimental conditions.

The biosorbent capacity and the removal percentage of Pb (II) were obtained as follows:

$$q = \frac{C_0 - C_f}{W} V \tag{1}$$

$$\% \, Removal = \frac{C_0 - C_f}{C_0} \cdot 100 \tag{2}$$

where q (mg g^{-1}) is the amount of adsorbed lead ions, C_0 (mg L^{-1}) is the initial concentration of lead, C_f (mg L^{-1}) is the final lead concentration, V (L) is the volume of the metal solution, and W (g) is the weight of the biosorbent.

2.4. Column Biosorption Experiments

Experiments in continuous systems, shown in Figure 1, were carried out using a glass column with a length of 3.7 cm and an internal diameter of 3.1 cm. The alginate-immobilized *Spirulina* mass contained in the column was of 10 g. The synthetic solutions containing Pb was circulated with a flow rate of 1.6 mL min^{-1} through the column using a peristaltic pump. Samples were collected at selected times until bed saturation.

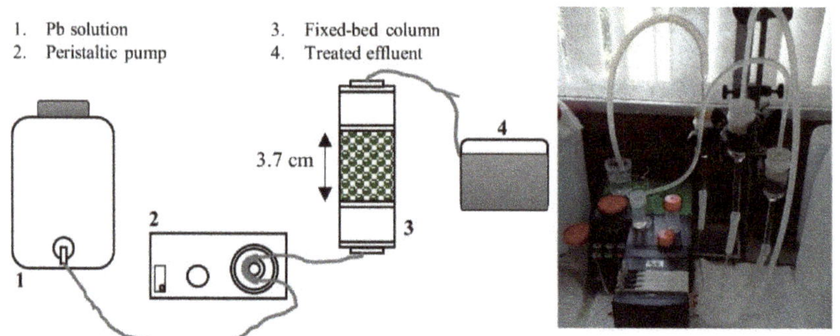

Figure 1. Setup of the experimental system.

The adsorption capacity was calculated from the breakthrough curve as:

$$q = \frac{C_0 Q}{1000\,W} \int_0^{t_e} \left(1 - \frac{C}{C_0}\right) dt \tag{3}$$

where C (mg L^{-1}) is the amount of metal ions in the treated solution, Q (mL min^{-1}) is the volumetric flow rate, W (g) is the dry weight of the biosorbent, and t_e (min) is the operation time of the column when $C/C_0 = 1$.

The initial concentration of Pb was varied from 4 to 20 mg L^{-1} to evaluate the influence of the metal concentration on column performance. The initial and final concentrations of Pb in aqueous solutions were determined by AAS (Varian SpectrAA-110).

2.5. Mathematical Models

2.5.1. Adsorption Isotherms

The Freundlich [31], Langmuir [32], and Dubinin-Radushkevich [33] isotherms (Equations (4)–(6), respectively) were fitted to equilibrium data by non-linear regression.

$$q_e = k_f\, C_e^{1/n} \tag{4}$$

where q_e (mg g^{-1}) is the amount of lead adsorbed per unit weight of biosorbent at equilibrium with a given solution concentration C_e (mg L^{-1}), k_f (mg$^{(n-1)/n}$ L$^{1/n}$ g^{-1}) is the adsorption capacity, and n (-) is a parameter related to the adsorption intensity.

$$q_e = \frac{q_{max\_L}(k_L\, C_e)}{1 + (k_L\, C_e)} \tag{5}$$

where q_{max_L} (mg g^{-1}) indicates the maximum amount of metal ions per unit mass of biomass at equilibrium, and k_L (L mg^{-1}) is the equilibrium adsorption constant.

$$q_e = q_{max\_DB}\, \exp\left(-\beta \varepsilon^2\right) \tag{6}$$

where q_{max_DB} (mg g^{-1}) is the maximum adsorption capacity, β (mol^2 J^{-2}) is a constant associated with the adsorption energy, and ε (J mol^{-1}) is Polanyi potential obtained as:

$$\varepsilon = RT\, ln\left(1 + \frac{1}{C_e}\right) \tag{7}$$

where R is the universal gas constant (8.314 J mol^{-1} K^{-1}).

The value of the mean adsorption energy, E (J mol^{-1}), is obtained as:

$$E = \frac{1}{\sqrt{2\beta}} \tag{8}$$

The value of E is widely used to make a prediction about the nature of the adsorption processes, either physical or chemical [34].

2.5.2. Sorption Dynamics in Fixed-Bed Columns

The understanding of the sorption phenomena in a fixed column is required for design and operation of a full-scale adsorption process. It should be noted that the description of the sorption in a continuous process is not straightforward due to the importance of axial dispersion, sorption kinetics, intraparticle diffusion resistance, and mass transfer. Therefore, several mathematical models have been reported to predict breakthrough behaviors and to calculate the design parameters for fixed-bed column adsorption.

The Thomas and Bohart-Adams models are frequently used to predict the dynamic behavior in fixed-column processes. Thomas' equation is obtained from a Langmuir adsorption equilibrium system with a pseudo second-order reaction kinetic law and without axial dispersion [35]:

$$\frac{C}{C_0} = \frac{1}{1 + \exp\left[\frac{k_{TH}}{Q}\left(q_{Th}W - C_0 V_{eff}\right)\right]} \tag{9}$$

where C and C_0 (mg L^{-1}) are the effluent and influent lead concentration, k_{TH} (mL mg^{-1} min^{-1}) is the Thomas rate constant, q_{TH} (mg g^{-1}) is the adsorption capacity, W (g) is the mass of biosorbent, Q (mL min^{-1}) is the flow rate, and V_{eff} (L) the volume of the effluent.

The Bohart-Adams model was founded on the surface rate theory and supposes that equilibrium is not achieved instantaneously. This model is usually applied to describe the initial part of the breakthrough curve [36]. The equation can be expressed as:

$$\frac{C}{C_0} = \exp\left(k_{AB}C_0 t - k_{AB}N_0\frac{Z}{v}\right) \tag{10}$$

where k_{AB} (L mg^{-1} min^{-1}) is the rate constant, N_0 (mg L^{-1}) is the saturation concentration, Z (cm) is the bed depth, and v (cm min^{-1}) is the linear velocity obtained by dividing the flow rate by the column section area.

The modified dose–response model [37] was also implemented to evaluate experimental values. This empirical model focuses on the description of the kinetics of metal removal in a biosorption column and minimizes the error generated by previous models, especially for lower and higher times of the breakthrough curve.

$$\frac{C}{C_0} = 1 - \frac{1}{1 + \left(\frac{V_{eff}}{b}\right)^a} \tag{11}$$

where a (-) and b (L) are the model parameters.

2.5.3. Mathematical Model Parameter Determination

With the aim of determining all model parameters, a non-linear square method was applied using Solver-add MS-Excel. The differences between values predicted by the model and the experimental values were estimated by the root mean square errors (*RMSEs*) as:

$$RMSE = \sqrt{\frac{\sum_1^m \left(X_{exp} - X_{model}\right)^2}{m}} \tag{12}$$

where m is the number of data points and X_{exp} and X_{model} are the experimental and the predicted values, respectively, of the adsorption capacity, q (mg g^{-1}), or the fractional concentration, C/C_0, for isotherm or fixed-bed adsorption models, respectively.

The minimization of *RMSE* values with simultaneous variation of model parameters allowed the fit of the experimental data to the non-linear forms of the equations. With the purpose of comparing the results obtained from different models, the coefficient of determination (R^2) of parity plots (i.e., experimental versus model values) was determined.

3. Results and Discussion

3.1. Effect of Initial pH of the Solution

The effect of the pH value of the aqueous solution on the biosorption of Pb (II) ions on *Spirulina* was evaluated. The changes of Pb (II) adsorption capacity at different initial pH values (from 2 to 5.2) are depicted in Figure 2. As can be seen, the Pb (II) biosorption capacity of *Spirulina* is strongly dependent on pH value. When the solution pH was below 3, the biosorption capacity was significantly lower. These results are in agreement with the competition between protons and Pb (II) for adsorption sites, and with the increase in electrostatic repulsion between positively charged species. At these pH values, Pb is mainly present as Pb^{2+} [38]. With increasing pH, the biosorption capacity of Pb (II) on *Spirulina* increased. These results have been previously related to the occurrence of more ligands carrying negative charges, such as carboxyl groups, which entails higher attraction of metal ions and biosorption onto biosorbent surfaces [39]. According to the experimental results, no significant differences in Pb (II) removal were observed for pH values from 4.5 to 5.2. As 5.2 was the pH of the original aqueous Pb (II) solution, this value was selected for further experiments. Previous studies dealing with the biosorption of Pb (II) have found similar optimal pH values [40–42].

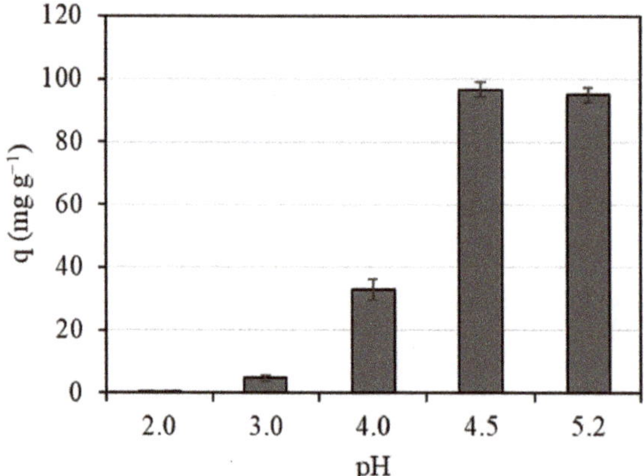

Figure 2. Effect of pH on adsorption capacity of Pb (II) by *Spirulina* (C_0 = 5.63 mg L^{-1}, S/L = 50 mg L^{-1}).

3.2. Effect of Ionic Strength on Pb (II) Biosorption

The influence of ionic strength on Pb (II) biosorption was evaluated by adding solutions of 0.01 and 0.1 M NaNO$_3$ at pH 5.2. The experimental results, presented in Figure 3, show the importance of the ionic strength on biosorption of lead by *Spirulina*. The adsorption capacity of Pb (II) was 95.07 mg g^{-1} for an initial Pb concentration of 5.63 mg L^{-1} and pH 5.2. The adsorption capacity decreased linearly by increasing ionic strength, obtaining 90.2 and 36.7 mg kg^{-1} at ionic strength 0.01 and 0.1 M, respectively, under the same experimental conditions.

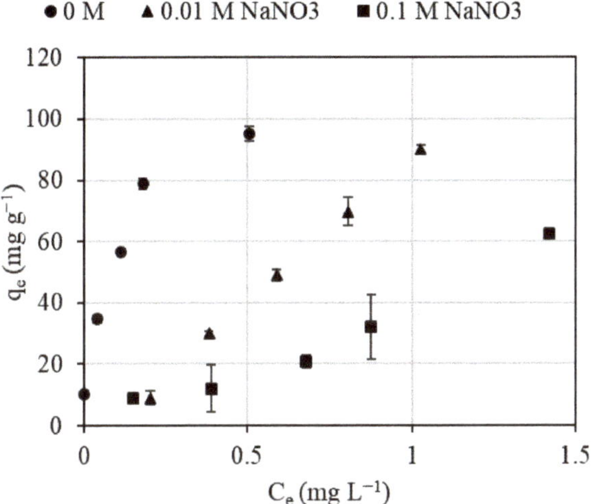

Figure 3. Evolution of adsorbed Pb (II) concentration on *Spirulina* for different ionic strengths (experimental conditions: *S/L*: 50 mg L^{-1}, contact time: 72 h, pH: 5.2).

Changes in the ionic strength not only affect the interfacial potential and, subsequently, the activity of electrolyte ions, but also the competition between the electrolyte ions. According to the experimental results, it could be concluded that Na ions displace Pb ions. Previous studies have explained the effect of ionic strength as an electrostatic competition of cations added as salts with heavy metals. In other words, for high ionic strength, adsorption sites are covered by counter-ions, which entail a loss of charge, and, consequently, this weakens the binding forces due to electrostatic interactions [39]. These results suggest that ion exchange is the most relevant mechanism for Pb (II) biosorption on *Spirulina*, as discussed in previous work carried out with other algae [43]. Other interactions, i.e., complexation/coordination, electrostatic interactions, chemisorption, physisorption, microprecipitation, and reduction, can also coexist with ion exchange [4].

3.3. Adsorption Isotherms

The Freundlich, Langmuir, and Dubinin-Radushkevich isotherm models were applied to study the equilibrium data. The experimental results were simulated with non-linear forms of these model equations, as presented in Figure 4. As can be seen in Table 1, these three models properly describe the biosorption process with high correlation coefficients and low *RSME* values. The Langmuir isotherm shows a slightly better fit with a correlation coefficient (R^2) of 0.992. Pb (II) biosorption can be regarded as more likely with monolayer adsorption, as concluded in previous studies dealing with biosorption of metals onto algae [44–46]. The maximum adsorption capacity according to the Langmuir model was 114.47 mg g^{-1}, which is comparable to the results from other studies (Table 2). The Dubinin-Radushkevich isotherm allows the description of the nature of the biosorption process through the estimation of the free energy of sorption. From the value of β, the value of the free energy obtained was higher than 8 kJ mol^{-1}, which has been associated with chemical adsorption of Pb on *Spirulina*'s surface [34]. These results are consistent with previous studies dealing with *Spirulina* as a biosorbent [18].

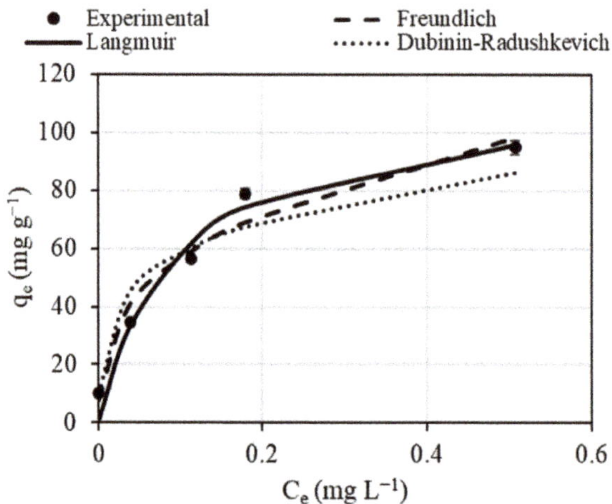

Figure 4. Equilibrium adsorption isotherm for the adsorption of Pb (II) on *Spirulina* (pH = 5.23 (not adjusted), S/L = 50 mg L^{-1}).

Table 1. Equilibrium isotherm parameters for biosorption of Pb (II) on *Spirulina*. *RMSE*, root mean square error.

Model	Parameters	
Freundlich	K_f (mg$^{(n-1)/n}$ L$^{1/n}$ g^{-1})	125.9
	$1/n$	0.35
	RMSE	7.23
	R^2	0.982
Langmuir	k_L (L mg^{-1})	10.25
	$q_{\text{max\_}L}$ (mg g^{-1})	114.47
	RMSE	3.09
	R^2	0.992
Dubinin-Radushkevich	$q_{\text{max\_}DR}$ (mg g^{-1})	378.15
	β (mol^2 kJ^{-2})	1.44×10^{-3}
	E (kJ mol^{-1})	18.6
	RMSE	9.23
	R^2	0.975

Table 2. Adsorption capacity of Pb (II) by different biosorbents.

Biosorbent Type	q (mg Pb g^{-1})	Reference
Polysiphonia	126.5	[47]
Cladophora glomerata	73.5	[47]
Gracilaria corticata	54	[47]
Polysiphonia violacea	102	[47]
Composted spent mushroom	125.2	[48]
Spirogyra sp.	140.84	[49]
Chlorella sp.	10.4	[5]
Rhizoclonium hookeri	81.7	[5]
Spirulina	114.47	(This work)

3.4. Immobilized Biomass

3.4.1. Isotherm Studies

Batch experiments were carried out with alginate gel and alginate-immobilized *Spirulina* to evaluate the isotherm models, i.e., Freundlich, Langmuir, and Dubinin-Radushkevich. The equilibrium isotherm parameters for the biosorption of Pb (II) under these experimental conditions are presented in Table 3. As can be seen, the three selected models fit properly to the experimental data. According to the Langmuir model, the maximum adsorption capacity was 303.94 and 282.17 mg g^{-1} for alginate and immobilized *Spirulina*, respectively. These values are consistent with other studies dealing with alginate as a matrix of a natural polymer to immobilize algal biomass [25]. The higher values obtained for biosorption capacity corroborate the potential of alginate as biosorbent in Pb (II) removal. Regarding the value of the free energy obtained from the Dubinin-Radushkevich parameters, it could be concluded that the adsorption of Pb is of a chemical nature in agreement with the results obtained for free *Spirulina*.

Table 3. Equilibrium isotherm parameters for the biosorption of Pb (II) on alginate and *Spirulina* immobilized in alginate.

Model	Parameters	Alginate	*Spirulina* Immobilized in Alginate
Freundlich	k_f	179.4	136.96
	$1/n$	0.147	0.21
	$RMSE$	22.27	22.97
	R^2	0.946	0.967
Langmuir	k_L (L mg^{-1})	1.54	1.58
	q_{max_L} (mg g^{-1})	303.94	282.17
	$RMSE$	19.92	20.53
	R^2	0.973	0.976
Dubinin-Radushkevich	q_{max_DR} (mg g^{-1})	571.99	689.32
	β (mol^2 kJ^{-2})	1.33×10^{-3}	1.78×10^{-3}
	E (kJ mol^{-1})	19.3	16.7
	$RMSE$	20.39	21.75
	R^2	0.957	0.974

3.4.2. Fixed-Bed Column Studies

The breakthrough curve obtained for a flow rate of 1.6 mL min^{-1} and for an inlet metal ion concentration of 20 mg L^{-1} is presented in Figure 5. Initially, most of the Pb (II) bound close to the inlet of the column, while the downstream sections were not exposed to the metal. The shape of this curve depends on several experimental factors, such as inlet flow rates, concentrations, and other properties of the column, i.e., diameter and bed height. The section of the bed in which the concentration of Pb (II) changes importantly is known as the mass transfer zone (MTZ). The shape of this section depends on the equilibrium equation, the axial dispersion in the reactor, and the kinetics of adsorption. A square-wave curve is the ideal shape of a breakthrough, since the column would allow the largest volume of contaminated water to be treated, and the maximal amount of metal would be removed during operation. Therefore, the shape of the breakthrough curve experimentally obtained could indicate that the equilibrium between the adsorbent and the surrounding solution was not reached instantaneously [50].

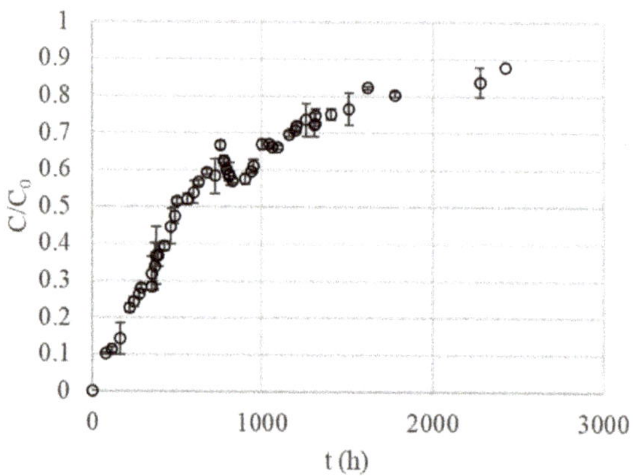

Figure 5. Breakthrough curve for the biosorption of Pb (II) on *Spirulina* (C_0 = 20 mg L^{-1}, Q = 1.6 mL min^{-1}, W = 10 g).

The mathematical description of this curve is required not only for designing a fixed-bed adsorber, but also to predict the behavior of the dynamic system. The estimated parameters obtained by non-linear regression analysis are presented in Table 4. Neither the Thomas nor the Bohart-Adams model adequately reproduced the experimental data, giving R^2 values of 0.907 and 0.849, respectively.

Table 4. Model parameters obtained by non-linear regression analysis for column studies.

Model	Parameters	Values
Thomas	k_{TH} (mL mg^{-1} min^{-1})	1.096×10^{-3}
	q_{TH} (mg g^{-1})	146.2
	RMSE	6.49×10^{-2}
	R^2	0.907
Bohart-Adams	k_{AB} (L mg^{-1} min^{-1})	3.726×10^{-4}
	N_0 (mg L^{-1})	157.69
	RMSE	8.16×10^{-2}
	R^2	0.849
Modified dose-response	a	1.153
	b (L)	64.94
	RMSE	3.88×10^{-2}
	R^2	0.967

As can be concluded from the *RSME* and R^2 values (Table 4), the dose-response model shows the best fit to the experimental data. These results are in accordance with previous studies that found that this model minimizes the most relevant errors obtained from other models, particularly at low and high values of the operation time [51].

The dependence of the adsorption capacity of immobilized *Spirulina* on alginate on the initial metal concentration of Pb (II) is shown in Figure 6. According to the experimental results, the biosorption capacity increased linearly with an increase in the initial Pb (II) concentration. As the system is characterized by a Langmuir isotherm, the maximum value of adsorption capacity could be determined from model parameters presented in Table 3. The maximum adsorption capacities (q^*) for initial concentrations of 4 and 20 mg L^{-1} were 243.91 and 273.51 mg g^{-1}, respectively, while the experimental adsorption capacity (q) was 47.98 and 202.85 mg g^{-1}, respectively. These results indicate that sorption was controlled by the mass transfer for the flow rate studied. In other words, the metal was not able to reach all positions of the biosorbent surface in a time shorter than residence time, in agreement with the conclusions from the breakthrough curve shape (Figure 5).

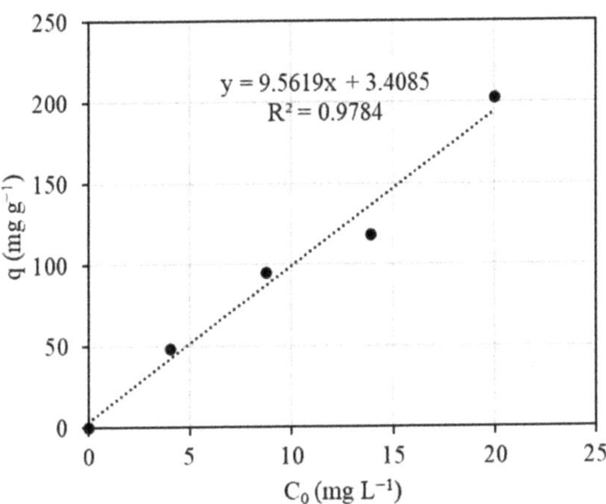

Figure 6. Dependence of the adsorption capacity of *Spirulina* on the initial metal concentration of Pb (II).

4. Conclusions

Immobilized *Spirulina* on alginate beads was proposed as an effective biosorbent for lead removal in batch and fixed-bed columns. From batch experiments, the biosorption capacity of *Spirulina* was concluded to be strongly dependent on pH value. The experimental results suggested that ion exchange is probably the most relevant mechanism for Pb (II) sorption on *Spirulina*. The Langmuir model properly described the isotherm data of lead biosorption by free and immobilized *Spirulina*, indicating a monolayer sorption onto the biomass surface. The potential of Pb (II) adsorption from *Spirulina* using immobilized biomass in a continuous fixed-bed column was assessed. The breakthrough curve was applied to study the efficiency of the column using immobilized *Spirulina*. The dose-response model described better the experimental data than both the Thomas and the Bohart-Adams models, which is associated with the minimization of relevant errors obtained at low and high values of the operation time. The importance of the influent metal concentration was also evaluated, concluding that the adsorption capacity of *Spirulina* increased linearly with the initial metal concentration. From the comparison of batch and column studies, it was found that equilibrium between *Spirulina* and the solution was not instantaneously reached in the dynamic studies. The promising results obtained for immobilized *Spirulina* on alginate beads could be the first step of the valorization of algal biomass for the treatment of industrial wastewater contaminated by toxic metals such as Pb.

Author Contributions: Conceptualization, all authors; methodology, C.J. and J.M.R.-M.; software, J.M.R.-M. and M.V.-G.; validation, C.J. and J.M.R.-M.; investigation, all authors; resources, all authors; data curation, all authors; writing—original draft preparation, M.V.-G.; writing—review and editing, all authors; supervision, C.J. and J.M.R.-M. All authors have read and agreed to the published version of the manuscript.

Funding: This research received no external funding.

Institutional Review Board Statement: Not applicable.

Informed Consent Statement: Not applicable.

Data Availability Statement: Not applicable.

Acknowledgments: The University of Malaga is acknowledged for the financial support from the "Plan Propio de Investigación".

Conflicts of Interest: The authors declare no conflict of interest.

References

1. Tchounwou, P.B.; Yedjou, C.G.; Patlolla, A.K.; Sutton, D.J. Heavy Metal Toxicity and the Environment. *Mol. Clin. Environ. Toxicol.* **2012**, *101*, 133–164. [CrossRef]
2. Environmental Protection Agency. *Lead and Copper Rule Revisions White Paper*; Environmental Protection Agency: Washington, DC, USA, 2016.
3. Beni, A.A.; Esmaeili, A. Biosorption, an Efficient Method for Removing Heavy Metals from Industrial Effluents: A Review. *Environ. Technol. Innov.* **2020**, *17*, 100503. [CrossRef]
4. Torres, E. Biosorption: A Review of the Latest Advances. *Processes* **2020**, *8*, 1584. [CrossRef]
5. Ubando, A.T.; Africa, A.D.M.; Maniquiz-Redillas, M.C.; Culaba, A.B.; Chen, W.-H.; Chang, J.-S. Microalgal Biosorption of Heavy Metals: A Comprehensive Bibliometric Review. *J. Hazard. Mater.* **2021**, *402*, 123431. [CrossRef]
6. Davis, T.A.; Volesky, B.; Mucci, A. A Review of the Biochemistry of Heavy Metal Biosorption by Brown Algae. *Water Res.* **2003**, *37*, 4311–4330. [CrossRef]
7. Brouers, F.; Al-Musawi, T.J. On the Optimal Use of Isotherm Models for the Characterization of Biosorption of Lead onto Algae. *J. Mol. Liq.* **2015**, *212*, 46–51. [CrossRef]
8. Matheickal, J.T.; Yu, Q. Biosorption of Lead from Aqueous Solutions by Marine Algae *Ecklonia radiata*. *Water Sci. Technol.* **1996**, *34*, 1–7. [CrossRef]
9. Areco, M.M.; Hanela, S.; Duran, J.; dos Santos Afonso, M. Biosorption of Cu(II), Zn(II), Cd(II) and Pb(II) by Dead Biomasses of Green Alga *Ulva lactuca* and the Development of a Sustainable Matrix for Adsorption Implementation. *J. Hazard. Mater.* **2012**, *213*, 123–132. [CrossRef]
10. Verma, A.; Kumar, S.; Kumar, S. Biosorption of Lead Ions from the Aqueous Solution by *Sargassum filipendula*: Equilibrium and Kinetic Studies. *J. Environ. Chem. Eng.* **2016**, *4*, 4587–4599. [CrossRef]
11. Li, Z.-Y.; Guo, S.-Y.; Li, L. Study on the Process, Thermodynamical Isotherm and Mechanism of Cr(III) Uptake by *Spirulina* Platensis. *J. Food Eng.* **2006**, *75*, 129–136. [CrossRef]
12. Rangabhashiyam, S.; Balasubramanian, P. Characteristics, Performances, Equilibrium and Kinetic Modeling Aspects of Heavy Metal Removal Using Algae. *Bioresour. Technol. Rep.* **2019**, *5*, 261–279. [CrossRef]
13. Zinicovscaia, I.; Safonov, A.; Zelenina, D.; Ershova, Y.; Boldyrev, K. Evaluation of Biosorption and Bioaccumulation Capacity of Cyanobacteria Arthrospira (*Spirulina*) Platensis for Radionuclides. *Algal Res.* **2020**, *51*, 102075. [CrossRef]
14. Lebron, Y.A.R.; Moreira, V.R.; Santos, L.V.S.; Jacob, R.S. Remediation of Methylene Blue from Aqueous Solution by Chlorella Pyrenoidosa and *Spirulina* Maxima Biosorption: Equilibrium, Kinetics, Thermodynamics and Optimization Studies. *J. Environ. Chem. Eng.* **2018**, *6*, 6680–6690. [CrossRef]
15. Gong, R.; Ding, Y.; Liu, H.; Chen, Q.; Liu, Z. Lead Biosorption and Desorption by Intact and Pretreated *Spirulina* Maxima Biomass. *Chemosphere* **2005**, *58*, 125–130. [CrossRef]
16. Chojnacka, K.; Chojnacki, A.; Górecka, H. Biosorption of Cr^{3+}, Cd^{2+} and Cu^{2+} Ions by Blue–Green Algae *Spirulina* Sp.: Kinetics, Equilibrium and the Mechanism of the Process. *Chemosphere* **2005**, *59*, 75–84. [CrossRef]
17. Solisio, C.; Lodi, A.; Torre, P.; Converti, A.; Del Borghi, M. Copper Removal by Dry and Re-Hydrated Biomass of *Spirulina* Platensis. *Bioresour. Technol.* **2006**, *97*, 1756–1760. [CrossRef]
18. Şeker, A.; Shahwan, T.; Eroğlu, A.E.; Yılmaz, S.; Demirel, Z.; Dalay, M.C. Equilibrium, Thermodynamic and Kinetic Studies for the Biosorption of Aqueous Lead(II), Cadmium(II) and Nickel(II) Ions on *Spirulina* Platensis. *J. Hazard. Mater.* **2008**, *154*, 973–980. [CrossRef] [PubMed]
19. Mahamadi, C. On the Dominance of Pb during Competitive Biosorption from Multi-Metal Systems: A Review. *Cogent Environ. Sci.* **2019**, *5*, 1635335. [CrossRef]
20. Almomani, F.; Bohsale, R.R. Bio-Sorption of Toxic Metals from Industrial Wastewater by Algae Strains *Spirulina* Platensis and Chlorella Vulgaris: Application of Isotherm, Kinetic Models and Process Optimization. *Sci. Total Environ.* **2020**, *755*, 142654. [CrossRef]
21. Sayadi, M.H.; Rashki, O.; Shahri, E. Application of Modified *Spirulina* Platensis and Chlorella Vulgaris Powder on the Adsorption of Heavy Metals from Aqueous Solutions. *J. Environ. Chem. Eng.* **2019**, *7*, 103169. [CrossRef]
22. Ramrakhiani, L.; Ghosh, S.; Majumdar, S. Surface Modification of Naturally Available Biomass for Enhancement of Heavy Metal Removal Efficiency, Upscaling Prospects, and Management Aspects of Spent Biosorbents: A Review. *Appl. Biochem. Biotechnol.* **2016**, *180*, 41–78. [CrossRef] [PubMed]
23. Iqbal, M.; Edyvean, R.G.J. Biosorption of Lead, Copper and Zinc Ions on Loofa Sponge Immobilized Biomass of Phanerochaete Chrysosporium. *Miner. Eng.* **2004**, *17*, 217–223. [CrossRef]
24. De Araujo, L.G.; de Borba, T.R.; de Pádua Ferreira, R.V.; Canevesi, R.L.S.; da Silva, E.A.; Dellamano, J.C.; Marumo, J.T. Use of Calcium Alginate Beads and Saccharomyces Cerevisiae for Biosorption of 241Am. *J. Environ. Radioact.* **2020**, *223*, 106399. [CrossRef]
25. Bayramoğlu, G.; Tuzun, I.; Celik, G.; Yilmaz, M.; Arica, M.Y. Biosorption of Mercury(II), Cadmium(II) and Lead(II) Ions from Aqueous System by Microalgae Chlamydomonas Reinhardtii Immobilized in Alginate Beads. *Int. J. Miner. Process.* **2006**, *81*, 35–43. [CrossRef]
26. McHale, A.P.; McHale, S. Microbial Biosorption of Metals: Potential in the Treatment of Metal Pollution. *Biotechnol. Adv.* **1994**, *12*, 647–652. [CrossRef]

27. Petrovič, A.; Simonič, M. Removal of Heavy Metal Ions from Drinking Water by Alginate-Immobilised Chlorella Sorokiniana. *Int. J. Environ. Sci. Technol.* **2016**, *13*, 1761–1780. [CrossRef]

28. Zarrouk, C. *Université de Paris Contribution à l'Étude d'une Cyanophycée: Influence de Divers Facteurs Physiques et Chimiques sur la Croissance et la Photosynthèse de Spirulina Maxima (Setch et Gardner) Geitler*; Faculté des Sciences de l'Université de Paris: Paris, France, 1966.

29. Lu, Y.; Wilkins, E. Heavy Metal Removal by Caustic-Treated Yeast Immobilized in Alginate. *J. Hazard. Mater.* **1996**, *49*, 165–179. [CrossRef]

30. OCDE. *Test. No. 106: Adsorption—Desorption Using a Batch Equilibrium Method*; OECD Publishing: Paris, France, 2000; ISBN 92-64-06960-7.

31. Freundlich, H.M. F Over the Adsorption in Solution. *J. Phys. Chem.* **1906**, *57*, 385–471.

32. Langmuir, I. The Constitution and Fundamental Properties of Solids and Liquids. Part I. Solids. *J. Am. Chem. Soc.* **1916**, *38*, 2221–2295. [CrossRef]

33. Dubinin, M.M. The Potential Theory of Adsorption of Gases and Vapors for Adsorbents with Energetically Nonuniform Surfaces. *Chem. Rev.* **1960**, *60*, 235–241. [CrossRef]

34. Hu, Q.; Zhang, Z. Application of Dubinin–Radushkevich Isotherm Model at the Solid/Solution Interface: A Theoretical Analysis. *J. Mol. Liq.* **2019**, *277*, 646–648. [CrossRef]

35. Thomas, H.C. Heterogeneous Ion Exchange in a Flowing System. *J. Am. Chem. Soc.* **1944**, *66*, 1664–1666. [CrossRef]

36. Bohart, G.S.; Adams, E.Q. Some Aspects of the Behavior of Charcoal with Respect to Chlorine. *J. Am. Chem. Soc.* **1920**, *42*, 523–544. [CrossRef]

37. Yan, G.; Viraraghavan, T.; Chen, M. A New Model for Heavy Metal Removal in a Biosorption Column. *Adsorpt. Sci. Technol.* **2001**, *19*, 25–43. [CrossRef]

38. Ford, R.G.; Wilkin, R.T.; Puls, R.W. *Monitored Natural Attenuation of Inorganic Contaminants in Ground Water*; Naiontal Risk Management Research Laboratory, U.S. Environmental Protection Agency: Washington, DC, USA, 2007; Volume 2.

39. Vilar, V.J.P.; Botelho, C.M.S.; Boaventura, R.A.R. Influence of PH, Ionic Strength and Temperature on Lead Biosorption by Gelidium and Agar Extraction Algal Waste. *Process. Biochem.* **2005**, *40*, 3267–3275. [CrossRef]

40. Ajitha, P.; Kumar, V.; Madhavan, S.; Gomathi, T.; Kannan, R.; Sudha, P.N.; Sukumaran, A. Removal of Toxic Heavy Metal Lead (II) Using Chitosan Oligosaccharide-Graft-Maleic Anhydride/Polyvinyl Alcohol/Silk Fibroin Composite. *Int. J. Biol. Macromol.* **2017**, *104*, 1469–1482. [CrossRef]

41. Nadeem, R.; Manzoor, Q.; Iqbal, M.; Nisar, J. Biosorption of Pb(II) onto Immobilized and Native Mangifera Indica Waste Biomass. *J. Ind. Eng. Chem.* **2016**, *35*, 185–194. [CrossRef]

42. Zhou, K.; Yang, Z.; Liu, Y.; Kong, X. Kinetics and Equilibrium Studies on Biosorption of Pb(II) from Aqueous Solution by a Novel Biosorbent: Cyclosorus Interruptus. *J. Environ. Chem. Eng.* **2015**, *3*, 2219–2228. [CrossRef]

43. Li, Y.-H.; Du, Q.; Peng, X.; Wang, D.; Wang, Z.; Xia, Y.; Wei, B. Physico-Chemical Characteristics and Lead Biosorption Properties of Enteromorpha Prolifera. *Colloids Surf. Biointerfaces* **2011**, *85*, 316–322. [CrossRef] [PubMed]

44. Rezaei, H. Biosorption of Chromium by Using *Spirulina* Sp. *Arab. J. Chem.* **2016**, *9*, 846–853. [CrossRef]

45. Solisio, C.; Lodi, A.; Soletto, D.; Converti, A. Cadmium Biosorption on *Spirulina* Platensis Biomass. *Bioresour. Technol.* **2008**, *99*, 5933–5937. [CrossRef]

46. Tavana, M.; Pahlavanzadeh, H.; Zarei, M.J. The Novel Usage of Dead Biomass of Green Algae of Schizomeris Leibleinii for Biosorption of Copper(II) from Aqueous Solutions: Equilibrium, Kinetics and Thermodynamics. *J. Environ. Chem. Eng.* **2020**, *8*, 104272. [CrossRef]

47. Jalali, R.; Ghafourian, H.; Asef, Y.; Davarpanah, S.J.; Sepehr, S. Removal and Recovery of Lead Using Nonliving Biomass of Marine Algae. *J. Hazard. Mater.* **2002**, *92*, 253–262. [CrossRef]

48. Liu, X.; Bai, X.; Dong, L.; Liang, J.; Jin, Y.; Wei, Y.; Li, Y.; Huang, S.; Qu, J. Composting Enhances the Removal of Lead Ions in Aqueous Solution by Spent Mushroom Substrate: Biosorption and Precipitation. *J. Clean. Prod.* **2018**, *200*, 1–11. [CrossRef]

49. Gupta, V.K.; Rastogi, A. Biosorption of Lead from Aqueous Solutions by Green Algae Spirogyra Species: Kinetics and Equilibrium Studies. *J. Hazard. Mater.* **2008**, *152*, 407–414. [CrossRef]

50. Benjamin, M.M.; Lawler, D.F. *Water Quality Engineering: Physical/Chemical Treatment Processes*; John Wiley & Sons: Hoboken, NJ, USA, 2013; ISBN 1-118-16965-4.

51. Calero, M.; Hernáinz, F.; Blázquez, G.; Tenorio, G.; Martín-Lara, M.A. Study of Cr (III) Biosorption in a Fixed-Bed Column. *J. Hazard. Mater.* **2009**, *171*, 886–893. [CrossRef]

processes

Article

Green Synthesis of A Novel MXene–CS Composite Applied in Treatment of Cr(VI) Contaminated Aqueous Solution

Hongyou Wan [1], Lan Nan [1,2], Huikai Geng [1], Wei Zhang [1,3,4,5,6,*] and Huanhuan Shi [1,*]

[1]   School of Ecology and Environment, Zhengzhou University, Zhengzhou 450001, China; hywan@zzu.edu.cn (H.W.); zzunanlan@163.com (L.N.); zzughuikai@163.com (H.G.)
[2]   Key Laboratory for Resources Utilization Technology of Unconventional Water of Gansu Province, Gansu Academy of Membrane Science and Technology Co., Ltd., Lanzhou 730020, China
[3]   Yellow River Institute for Ecological Protection & Regional Coordinated Development, Zhengzhou University, Zhengzhou 450001, China
[4]   Henan International Joint Laboratory of Water Cycle Simulation and Environmental Protection, Zhengzhou University, Zhengzhou 450001, China
[5]   Zhengzhou Key Laboratory of Water Resource and Environment, Zhengzhou University, Zhengzhou 450001, China
[6]   Henan Province Key Laboratory of Water Pollution Control and Rehabilitation Technology, Pingdingshan 467000, China
*   Correspondence: zhangwei88@zzu.edu.cn (W.Z.); hhshi@zzu.edu.cn (H.S.)

Citation: Wan, H.; Nan, L.; Geng, H.; Zhang, W.; Shi, H. Green Synthesis of A Novel MXene–CS Composite Applied in Treatment of Cr(VI) Contaminated Aqueous Solution. *Processes* **2021**, *9*, 524. https://doi.org/10.3390/pr9030524

Academic Editor: Jose Enrique Torres Vaamonde

Received: 6 February 2021
Accepted: 6 March 2021
Published: 14 March 2021

Abstract: The considerable amount of Cr(VI) pollutants in the aqueous environment is a significant environmental concern that cannot be ignored. A series of novel Mxene–CS inorganic–organic composite nanomaterials synthesized by using the solution reaction method was applied to treat the Cr(VI) contaminated water. The Mxene–CS composites were characterized through SEM (scanning electron microscope), XRD (X–ray diffraction), XPS (X–ray photoelectron spectroscopy), and FTIR (Fourier transform infrared). The XRD patterns (observed at 2θ of 18.1°, 35.8°, 41.5°, and 60.1°) and the FT–IR spectra (-NH$_2$ group for 1635 and 1517 cm^{-1}, and -OH group for 3482 cm^{-1}) illustrated that CS was successfully loaded on the Mxene. The effects of solution pH, the dosage of Mxene–CS, and duration time on the adsorption of Cr(VI) by synthesized Mxene–CS were investigated. The removal efficiency of Cr(VI) was increased from 12.9% to 40.5% with Mxene–CS dosage ranging from 0.02 to 0.12 g/L. The adsorption process could be well fitted by the pseudo–second–order kinetics model, indicating chemisorption occurred. The Langmuir isotherm model could be better to describe the process with a maximum adsorption capacity of 43.1 mg/g. The prepared novel Mxene–CS composite was considered as an alternative for adsorption of heavy metals from wastewater.

Keywords: Mxene–CS composite; Cr(VI) contaminated aqueous solution; Green Synthesis; treatment of wastewater; heavy metals

1. Introduction

Currently, with more rapid development of industrialization, more heavy metals in wastewater are derived from various industrial processes, such as mining, smelting, electroplating, tanning, and electrochemistry [1,2]. The heavy metals in wastewater would inevitably cause serious damage to the surface water, groundwater, and soil without proper treatment [2]. Among the reported heavy metals in wastewater, Cr(VI) is one of the most frequently detected in the natural environment, which was mainly derived from the chromium salt production and consumption process. The main existing oxidation states of Cr element in aqueous solution were Cr(III) and Cr(VI). The Cr(III) is a necessary trace element in our human body that is less harmful to the human body, compared with Cr(VI) [3,4]. Thus, more attention has been given to the treatment of Cr(VI)–polluted water/soil environment due to its potential threat to the ecological environment and human health [5]. Several methods have been proposed and investigated to remove Cr(VI) from the water environment, including the adsorption method [6–9], reduction method [10], electrolysis method [11], ion exchange method [12], and membrane separation method [13]. Among the above–mentioned methods, the adsorption route has many inherent advantages for its lower energy consumption and relatively higher removal performance.

Chitosan (CS) is a kind of natural high molecular compound generating in the deacetylation of chitin. The CS is colorless and odorless, environmentally friendly with relatively higher adsorption properties [14] and stronger flocculation effect [15], that has already been widely applied in the chemical environmental protection [16,17], food hygiene [18,19], biomedicine [20], papermaking [21], and textile industries [22]. What is more, the hydroxyl group or amino group in the CS molecular would be beneficial to remove other pollutants for it can form into a hydrogen bond with a dye molecule (containing a -N=N- structure) or combine with the heavy metal ions under certain conditions [23]. However, the CS is not structurally stable under the acidic pH condition for it would degrade slowly in the acid solution, resulting in a decrease in the viscosity and molecular weight of the CS [24]. Therefore, the application of CS in the wastewater is obviously restricted for its unstable structure, while we should pay more attention to improve the stability of the CS and correspondingly increase its removal performance of pollutants from wastewater [25]. Several routes, including carboxymethylation modification [26], alkylation modification [27], and quaternization modification [27], etc., have been introduced to modify the structure of CS. A novel magnetic CS composite adsorbent was synthesized by the sol–gel method, proving its adsorption capacity of Cu^{2+} by the composite adsorbent could reach up to 216.6 mg/g [28]. The three–dimensional graphene oxide–CS composite was prepared and applied in adsorption of Uranium(VI) with an adsorption capacity of 384.6 mg/g at a pH of 8.3 [29].

The transition metal carbide or nitride (Mxene) is a kind of two–dimensional (2D) graphene–like nanomaterial newly developed in recent years [30]. Generally, the Mxene is mainly referred to as the transition metal carbides, nitrides, or borides, with the formula of $M_{n+1}X_nT_x$. In the formula, M is presenting the transition metal, X presents the non–metallic, including elements of carbon, nitrogen, and boron, T is presenting the functional groups (including -F, -O, and -OH) [31–33]. The novel Mxene phase is obtained by exfoliating the "A" component of the MAX phase in the HF solution, while the A in the MAX is presenting the elements for the Group IIIA or IVA, and X refers to the element of C or N [34]. The Mxene has been used as the adsorbent for the removal of heavy metal ions [35,36] or organic molecules [37] from the aqueous solution. The Mxene material is of relatively high structural stability, which should support and improve the structural stability of the CS [33]. It should be reasonable to expect the composite of Mxene and CS should be an ideal choice for the treatment of Cr(VI) contaminated water.

Herein, we synthesize and report a novel composite material of Mxene–CS. To the best of our knowledge, this is the first attempt to synthesize the Mxene–CS composite materials and used in the removal of Cr(VI) from the aqueous solution. We have studied the effects of pH, the Mxene–CS dosage, and the reaction time on the adsorption performance of Cr(VI) by synthesized Mxene–CS.

2. Experimental Section

2.1. Materials

The raw materials of Ti_3AlC_2 powders (One of MAX for the preparation of Mxene) were purchased from Ningbo Jinlei Nano Material Technology Co., Ltd., Zhejiang, China. The CS ($\geq$95 wt % purity) was brought from Zhengzhou Paini Chemical reagent Co., Ltd., Henan, China. The HF was purchased from Luoyang Chemical Reagent Factory, Henan, China. The acetic acid was supplied by Tianjin Sailboat Chemical Reagent Technology Co., Ltd., Tianjin, China.

2.2. Synthesis of the Mxene–CS Composite

The novel Mxene–CS composite was prepared by the solution synthesis route. Firstly, the raw materials of Ti_3AlC_2 powders were passed through 400 mesh and then dissolved into the HF solutions (40% w/w water solution) in a PTFE beaker to obtain the mixed slurry (note that the HF is a dangerous inorganic acid, which would easily corrode the experimenter or the labware material). Then, the slurry was stirred in a magnetic stirrer (250 rpm) under room temperature for 96 h, followed by withdrawn and centrifugal filtration with deionized water until the pH of the suspension was neutral. The black liquid containing MXene was dried at 60 °C in a vacuum drying oven to obtain the dry MXene. The CS (0.82 g) was added into the 150 mL of acetic acid (5% w/w water solution) to obtain the mixed slurry, while the Mxene (0.82 g) was added into the mixed slurry and ultrasonically dispersed for 30 min. The obtained gelatinous product was withdrawn and placed in a vacuum oven at 60 °C to get the MXene–CS. Finally, the prepared MXene–CS composites were ground and passed through 20–60 mesh sieves, and stored for the adsorption experiments.

2.3. Material Characterization

The surface morphology of prepared MXene–CS composite was measured by scanning electron microscopy (SEM, Zeiss AURIGA CrossBeam Focused Ion Beam Electron Microscope, Baden–Württemberg, Germany). The powder XRD patterns of the prepared composite were determined by using an X–ray diffractometer (Empyrean, Panaco, Almelo, The Netherlands). The FT–IR spectra of CS, Mxene and the synthesized Mxene–CS were determined by using the spectrometer (TENSOR27, Bruker, Karlsruhe, Germany). The chemical states for the main elements, including Cr, Ti, and C, were measured by XPS analysis (Thermo Scientific Escalab 250Xi, New York, NY, USA). The Zeta potential of the Mxene–CS in the aqueous solution at the pH ranging from 2 to 12 was measured by using a Zeta potentiometer (JS94H, Shanghai Zhongchen Digital Technology Equipment Co., Ltd., ShangHai, China).

2.4. Adsorption of Cr(VI) by Synthesized Mxene–CS

The stock solution of 1000 mM Cr(VI) was prepared by dissolving a certain amount of $K_2Cr_2O_7$ into the deionized water, then diluted into certain concentrations for the batch experiment. During the batch adsorption experiments, the $K_2Cr_2O_7$ solution (100 mL) was mixed with the Mxene–CS in a 250 mL Erlenmeyer flask, and oscillated in a horizontal shaking bath to control the temperature. The concentrations of Cr(VI) in the solution before and after the adsorption process were determined by using the diphenyl carbohydrazide spectrophotometric method at a wavelength of 540 nm [38]. The effect of reaction parameters, including the Mxene–CS dosage (0.02, 0.04, 0.06, 0.08, 0.1, and 0.12 g/L), the contact time (30, 60, 120, 180, 240, 300, 360, 480, 720, 960, 1200, and 1440 min), the temperature (25, 35, and 45 °C), and initial pH (2, 3, 4, 5, 6, and 7) on the removal efficiency and capacity of Cr(VI) by Mxene–CS composites were measured, respectively.

$$\text{Removal efficiency} = \frac{X_0 - X_e}{X_0} \times 100\% \tag{1}$$

$$q = \frac{(X_0 - X_e) \times V}{m} \tag{2}$$

where X_0 (mg/L) and X_e (mg/L) are presenting the original concentration and equilibrium concentration of Cr(VI) in the solution, respectively, V (L) is the volume of the potassium bichromate solution, and m (g) is the mass of Mxene–CS used in the batch experiment.

2.4.1. Adsorption Kinetic Model

The adsorption kinetic model was adopted to investigate the adsorption control mechanism of Cr(VI) by Mxene–CS [39]. The adsorption experiments were carried out in the

Cr(VI) solution (50 mg/L), then further fit the results by using pseudo–first–order kinetic model and pseudo–second–order model as described by Equations (3) and (4) [40,41].

$$\ln(q_e - q_t) = \ln q_e - k_1 t \tag{3}$$

$$\frac{t}{q_t} = \frac{1}{k_2 q_e^2} + \frac{t}{q_e} \tag{4}$$

where q_e is adsorption capacity at equilibrium (mg/g), q_t is the sorption capacity at a certain reaction time (mg/g), t is the reaction time (min), k_1 is the first–order rate constant (min^{-1}), and k_2 is the second–order rate constant (g/mg min^{-1}).

2.4.2. Adsorption Isotherm

In addition to the adsorption rate, the adsorption capacity is also an important index for the adsorbent's large–scale application in the industry. The removal data of Cr(VI) by Mxene–CS were fitted by using two typical isothermal adsorption models (Langmuir and Freundlich) to analyze the experiment data as shown in Equations (5) and (6) [2].

$$\frac{X_e}{q_e} = \frac{X_e}{q_{max}} + \frac{1}{q_{max}b} \tag{5}$$

$$\ln q_e = \ln K + \frac{1}{n}\ln X_e \tag{6}$$

where X_e is presenting the equilibrium concentration of Cr(VI) in the solution (mg/L), q_e is the adsorption capacity at equilibrium (mg/g), q_{max} is the maximum adsorption capacity of Mxene–CS (mg/g), and b is a constant related to the adsorption energy. In Equation (6), n and K are the Freundlich empirical constants relating to the adsorption strength and the sorption capacity of the adsorbent, respectively.

2.4.3. Adsorption Thermodynamics

The adsorption thermodynamics was explored to make clear whether the adsorption process is exothermic or endothermic, while the Gibbs free energy change, entropy change, and enthalpy change of the adsorption process are calculated according to the thermodynamic formula as followed Equations (7)–(9).

$$\Delta G^0 = -RT \ln K_c \tag{7}$$

$$\ln K_c = \frac{\Delta S^0}{R} - \frac{\Delta H^0}{RT} \tag{8}$$

$$\Delta G^0 = \Delta H^0 - T\Delta S^0 \tag{9}$$

where R is the gas constant (8.314 J/(mol K)), T is the absolute temperature (K), and K_c is the thermodynamic equilibrium constant. A linear fit is performed by plotting $\ln K$ versus $1/T$, and ΔH^0 and ΔS^0 can be obtained from the slope and intercept of the fitted curve.

3. Results and Discussion

3.1. Characterization of the Synthesized Mxene–CS

The Mxene was firstly prepared through etching of Ti_3AlC_2 by HF solution, while the appearance and schematic diagram from Ti_3AlC_2 to Mxene are shown in Figure 1a. The size of the prepared Mxene particles would get smaller, indicating the structure for the original Ti_3AlC_2 was partly changed. The XRD patterns (Figure 1b) were illustrating that the characteristic peak at 2θ of 39° could not be observed, indicating that the Al atoms were successfully etched out [42].

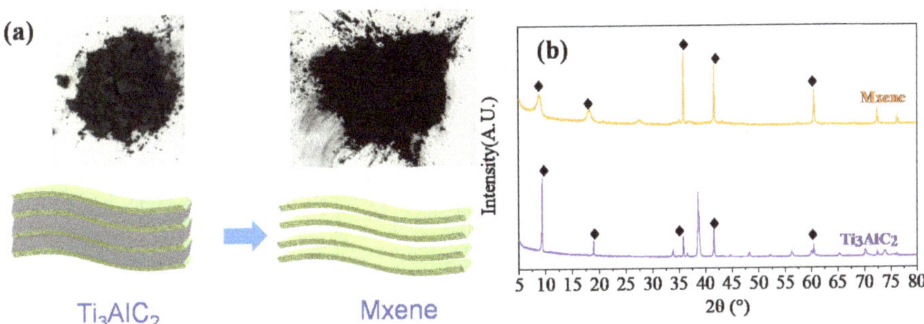

Figure 1. The preparation of Mxene by etching of Ti$_3$AlC$_2$: (**a**) The appearance and schematic diagram, (**b**) the comparison of XRD patterns between the Ti$_3$AlC$_2$ and Mxene.

The SEM images of original Ti$_3$AlC$_2$, Mxene, CS, and prepared Mxene–CS were shown in Figure 2. The pristine CS has a relative rough surface and compact structure, while the Mxene (after treatment of Ti$_3$AlC$_2$ powders by HF solutions) had a smoother surface and a typical two–dimensional layered structure, indicating that the Mxene was successfully prepared from the Ti$_3$AlC$_2$. For the Mxene–CS, the layered structure of Mxene does not completely disappear, and the irregular compact particles observed in the SEM image, indicating that CS was successfully loaded on the Mxene. The generation of the uneven surface and the pore structure in the Mxene–CS should enhance its specific surface area, which is beneficial to the contact between the surface–active group of Mxene–CS and the Cr(VI) pollutants in the aqueous solution. The XRD patterns and FT–IR spectra of Mxene, CS, and prepared Mxene–CS are shown in Figure 3. In the prepared Mxene–CS, the observed diffraction peaks observed at 2θ of 18.1°, 35.8°, 41.5°, and 60.1° were related to the Mxene. The rougher baseline for XRD of Mxene–CS also illustrated that the CS was entering into the Mxene. For the FT–IR spectra, the characteristic peaks of -NH$_2$ group (observed at 1635 and 1517 cm^{-1}) and -OH group (at 3482 cm^{-1}) in the CS are also presented in the Mxene–CS, indicating that the CS has been successfully entered into the prepared composite.

Figure 2. SEM images of the original Ti$_3$AlC$_2$ (**a**), the Mxene (**b**), CS (**c**), and prepared Mxene–CS (**d**).

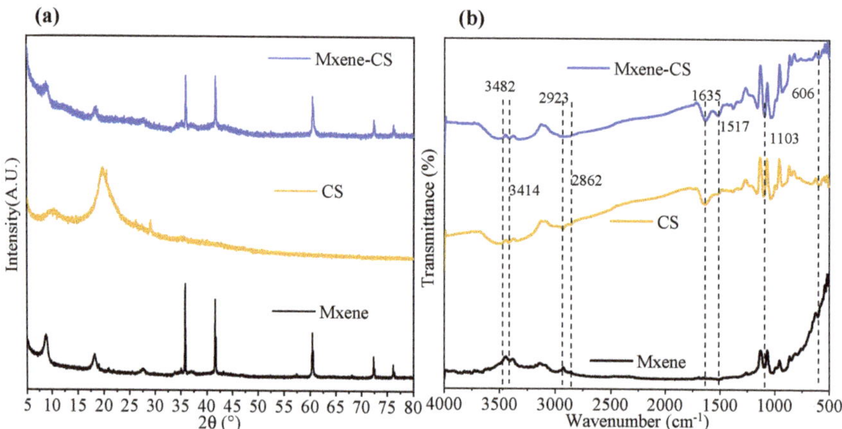

Figure 3. (a) XRD patterns of Mxene, CS, and prepared Mxene–CS. (b) FT–IR spectra of Mxene, CS, and prepared Mxene–CS.

3.2. Removal of Cr(VI) by Mxene–CS

3.2.1. Effect of Mxene–CS Dosage

The removal efficiency and capacity of Cr(VI) (50 mg/L) by the prepared Mxene–CS was shown in Figure 4. As shown, the removal efficiency of Cr(VI) was increased from 12.9% to 40.5% with Mxene–CS dosage ranging from 0.02 to 0.12 g/L, while the removal capacity was decreased from 30.5 to 16.1 mg/g. The progress of Cr(VI) removal efficiency was mainly attributed to more contact surface area and adsorption sites. However, the amount of Cr(VI) (which could be absorbed and removed) is limited, with more Mxene–CS in the aqueous solution, the amount of Cr(VI) adsorbed by per amount of Mxene–CS would be decreased as shown in Figure 4.

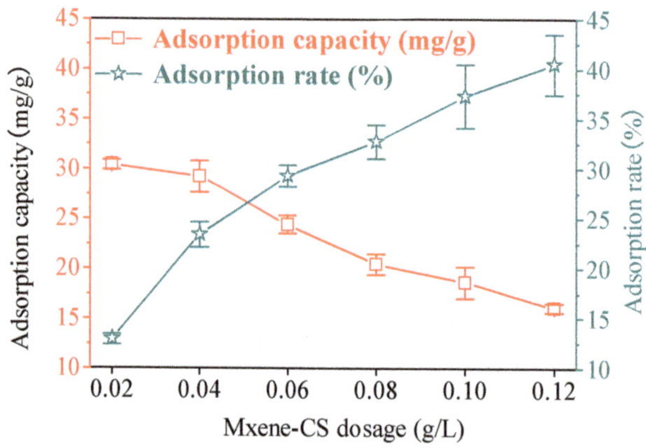

Figure 4. Effect of Mxene–CS dosage on the adsorption capacity and rate of Cr(VI) (Conditions: Adsorption temperature of 25 °C, pH of 4.3, and reaction time of 24 h).

3.2.2. Adsorption Kinetics Study

The adsorption kinetics results of Cr(VI) at varying initial Cr(VI) concentrations were shown in Figure 5, indicating that the removal efficiency increased within the initial 400 min and reached the equilibrium within 480 min at Mxene–CS dosage of 100 mg/L. The adsorption capacity (at equilibrium point) was calculated as 50.6, 13.0, and 10.10 mg/g at the initial Cr(VI) concentration of 100, 50, and 20 mg/L, respectively. It is obvious that the adsorption capacity is increasing slower after 480 min for the adsorption capacity of Mxene–CS reached saturation. Thus, the optimum reaction time was determined as 480 min.

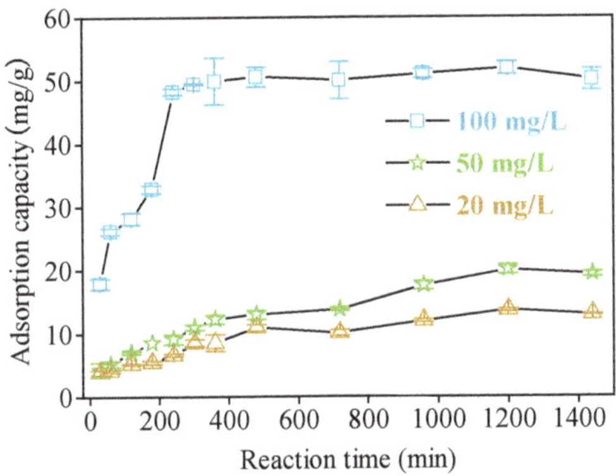

Figure 5. The adsorption capacity of Cr(VI) by Mxene–CS (0.6 g/L) with varying initial Cr(VI) concentration (Conditions: Adsorption temperature of 25 °C and pH of 4.3).

In order to further investigate the adsorption kinetics of Cr(VI) by Mxene–CS, the adsorption data were fitted with the pseudo–first–order kinetic model and the pseudo–second–order kinetic model (Table 1). The fitting kinetic curves were presented in Figure 6, while the higher R^2 value (0.9952) suggests that the removal of Cr(VI) by Mxene–CS is more prone to the pseudo–second–order kinetic model, further indicating that this removal process should be controlled by the chemical adsorption.

Table 1. The parameters for kinetic models of Cr(VI) removed by Mxene–CS.

Concentration (mg/L)	Pseudo–First–Order Model			Pseudo–Second–Order Model		
	q_e (mg/g)	K_1 (1/min)	R^2	q_e (mg/g)	K_2 (g/mg·min)	R^2
100	15.62	0.0032	0.5942	53.48	0.0003	0.9952
50	24.03	0.0053	0.5975	22.54	0.0002	0.9633
20	12.91	0.0028	0.6899	14.60	0.0002	0.9530

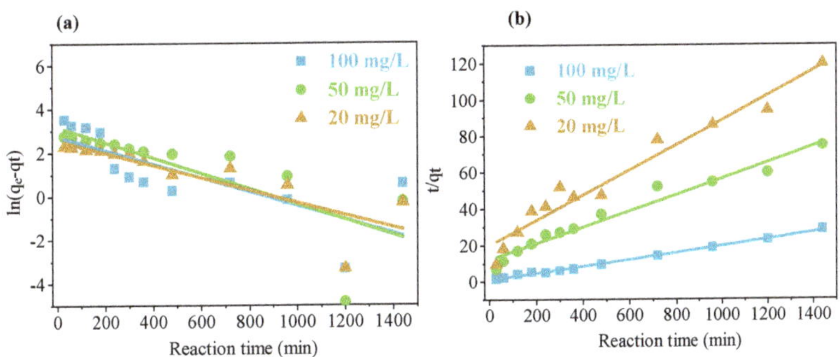

Figure 6. Fitting model for the Cr(VI) adsorption: The pseudo–first–order model (**a**), and the pseudo–second–order model (**b**).

3.2.3. Adsorption Isotherm Study

To further evaluate the removal capacity of Cr(VI) on Mxene–CS, the adsorption isotherm of Cr(VI) at different experimental conditions was investigated and is shown in Figure 7, while the corresponding fitting parameters are shown in Table 2. The Langmuir isotherm (with R^2 value of 0.9693 at 298 K) was observed to better fit the adsorption behaviors of Cr(VI) by Mxene–CS, indicating that the adsorption process is mainly monolayer adsorption. This is also consistent with the chemisorption process proved by pseudo–second–order kinetic model from the adsorption kinetics study.

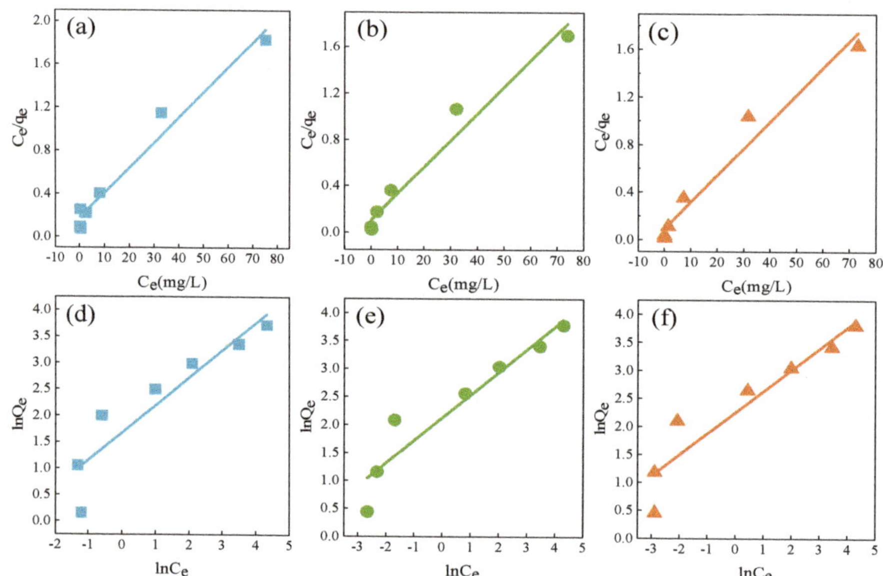

Figure 7. The Langmuir isotherm model curve for Mxene–CS (**a–c**) and the Freundlich isotherm model curve (**d–f**).

Table 2. The parameters and correlation coefficients for the Langmuir and Freundlich isotherm model.

T(K)	Langmuir Parameters			Freundlich Parameters		
	q_{max} (mg/g)	K_L (L/mg)	R^2	$1/n$	K_f	R^2
298	43.10	0.1305	0.9693	0.5205	e^1.6761	0.8576
308	43.48	0.2005	0.9659	0.4058	e^2.1217	0.9058
318	44.25	0.2545	0.9631	0.3794	e^2.2497	0.8910

3.2.4. Adsorption Thermodynamics Study

The adsorption thermodynamics of Cr(VI) by the prepared Mxene–CS were also studied and are shown in Figure 8 and Table 3. The adsorption free energy (ΔG^0) of negative values indicated that the adsorption of Cr(VI) by Mxene–CS is a spontaneous process. The enthalpy change value (ΔH^0) is more than zero, illustrating that the adsorption process is an endothermic reaction, which could explain the phenomenon that the adsorption efficiency of Cr(VI) was increased with increasing temperature. ΔS^0 is the algebraic sum of the entropy change of the reaction system, representing the degree of chaos. The adsorption entropy change (ΔS^0) is also more than zero, indicating that the obtained entropy (caused by desorption of water molecules) would exceed the entropy lost by the adsorption process.

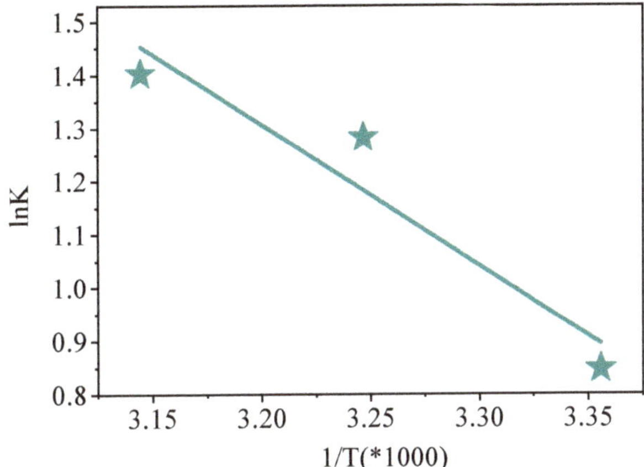

Figure 8. Thermodynamic fitting curve of adsorption of Cr(VI) on Mxene–CS.

Table 3. Thermodynamic parameter values for Cr(VI) adsorption.

T	ΔG^0 (kJ/mol)	ΔS^0 (J/mol)	ΔH^0 (kJ/mol)
298	−0.2522	22.0146	81.3084
308	−0.3820		
318	−0.4177		

3.3. Effect of pH

As pH values are variable in the actual water environment, the adsorption rate of Cr(VI) by Mxene–CS at pH ranging from 2 to 7 was investigated and is shown in Figure 9a. As shown, the pH would affect the adsorption capacity of Cr(VI) by the Mxene–CS, with the maximum capacity of 22.3 mg/g at pH of 4. To study the effects on Mxene–CS, we measured the zeta potential of the composite at different pH. The effect of pH on the zeta potential of Mxene–CS is shown in Figure 9b. The surface charge of Mxene–CS is all positive at pH ranging from 2 to 10, with the maximum value at pH around 4–6. Thus, at pH 2–7, the Mxene–CS surface is positively charged and has an electrostatic attraction between the composite and the negatively charged $Cr_2O_7^{2-}$. The decrease of adsorption capacity at pH ranging from 4 to 7 reflects the reduction in the zeta potential [5].

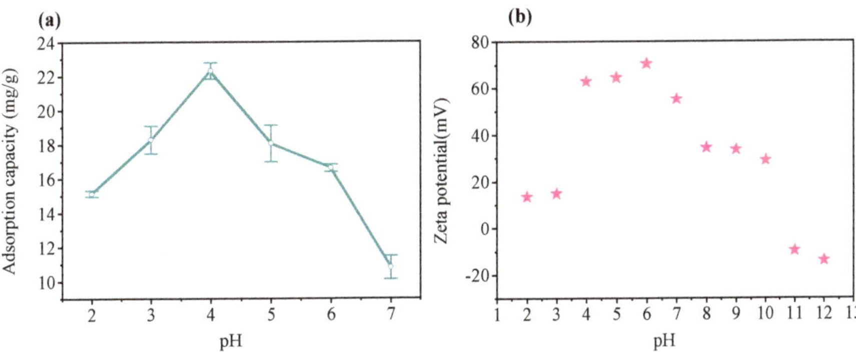

Figure 9. Effect of solution pH on adsorption performance of Cr(VI): (**a**) The influence of pH on adsorption capacity (Conditions: Temperature of 25 °C, and reaction time of 24 h), (**b**) the Zeta potential of Mxene–CS in the aqueous solution.

The XPS spectra for chemical states of Ti, C, Cr, and O were measured before and after the adsorption process, as shown in Figure 10a. It was found that a Cr 2p peak was observed at near 577.4 eV after Cr(VI) adsorption, indicating adsorption of Cr(VI) has occurred [43]. The XPS spectra of the Cr 2p core level are shown in Figure 10b, while the Cr 2p1/2 and Cr 2p3/2 peaks near 586.68 eV and 577.4 eV were detected, indicating that Cr(VI) was contacted and removed to the surface of the composite [43]. The comparison for removal of Cr(VI) by different materials from an aqueous solution is shown in Table 4. Compared with the prepared Chitosan–FeO nanoparticles, nZVI–multiwalled carbon nanotube, ZVMg, and modified brown algae, Sargassum bevanom, the removal performance of Cr(VI) by the Mxene–CS in this study is partly restricted, probably attributed to the relatively weak combination between the Mxene and CS at the certain molar ratio [24]. To expand its practical application in the actual wastewater, varying molar ratios of Mxene to CS on synthesizing of the novel composite are recommendable.

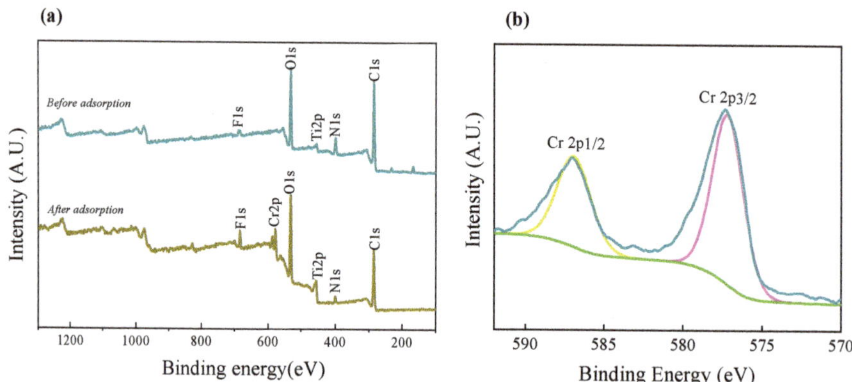

Figure 10. XPS spectra from Mxene–CS before and after adsorption: (**a**) Full–range XPS spectra, (**b**) Cr 2p.

Table 4. Comparasion for removal of Cr(VI) through different materials from an aqeous solution.

Order	Adsorbent	Optimum Condition (°C, pH, and the Dosage of Materials)	Initial Cr Concentration (mg/L)	Removal Efficiency (%)	Removal Capacity (mg/g)	Cited Reference
1	Chitosan–FeO nanoparticles	20, 6.0, 0.4 g/L	70	—	60.2 (the theoretical removal capacity)	[44]
2	nZVI–multiwalled carbon nanotube	30, 7.0, 0.1 g/L	20	98%	—	[45]
3	bentonite–supported nanoscale zero–valent iron	30, 5.0, 4 mg/L	20	99.8%	—	[46]
4	ZVMg	21, 7.0, 0.05 g/L	18.6	100%	—	[47]
5	ZVMg/combined with ultrasonic power treatment (100 W)	20–25, >7, (2.5 g/L, 5 g/L, and 10 g/L)	50	100%	—	[48]
6	ZVI/MgO	20, 5.0, 16 g/L	30	97%	—	[49]
7	Modified brown algae Sargassum bevanom	Room temperature, 3, 7 g/L	100	89.64%	—	[9]
8	Mxene–CS (This study)	25, 4.3, 0.6 g/L	100	30.4%	50.6 mg/g	This study

— presenting "not mentioned".

4. Conclusions

In this study, the Mxene–CS composite was successfully prepared by the simple solution method and applied in the removal of Cr(VI) from the aqueous solution. The following conclusions can be drawn from the batch experiment.

(1) In the prepared Mxene–CS, the layered structure of Mxene does not completely disappear, with the irregular compact particles detected, indicating that CS was successfully loaded on the Mxene.

(2) The effects of parameters, including Mxene–CS dosage and pH, on the removal of Cr(VI) from the aqueous solution were investigated. With the Mxene–CS dosage ranging from 0.02 to 0.12 g/L, the removal efficiency of Cr(VI) was increased from 12.9% to 40.5%. The influence of pH on the removal of Cr(VI) by the Mxene–CS was mainly controlled by the varying potential at different pH.

(3) The removal of Cr(VI) by Mxene–CS is more prone to the pseudo–second–order kinetic model, indicating the removal process should be controlled by the chemical adsorption. The Langmuir isotherm was more suitable to describe the adsorption behaviors of Cr(VI) by Mxene–CS, with the maximum adsorption capacity of 43.1 mg/g.

This experiment should provide a reliable theoretical basis for Mxene–CS adsorption of Cr(VI), and further expand the application prospect of inorganic–organic composites in wastewater.

Author Contributions: H.W.: Writing—Original draft preparation; L.N.: Investigation; H.G.: Investigation; W.Z.: Writing—Reviewing and Editing; H.S.: Supervision. All authors have read and agreed to the published version of the manuscript.

Funding: This research was funded by the Natural Sciences Foundation of China [52000163], the Natural Science Foundation of Henan Province [202300410423], and the China Postdoctoral Science Foundation [2020M682354].

Institutional Review Board Statement: Not applicable.

Informed Consent Statement: Not applicable.

Data Availability Statement: Not applicable.

Acknowledgments: The author would also thank the support from the Modern Analytical Computing Center of Zhengzhou University.

Conflicts of Interest: The authors declare no conflict of interest.

References

1. Strzebońska, M.; Jarosz-Krzemińska, E.; Adamiec, E. Assessing historical mining and smelting effects on heavy metal pollution of river systems over span of two decades. *Water Air Soil Pollut.* **2017**, *228*, 141. [CrossRef]
2. Wang, R.-Y.; Zhang, W.; Zhang, L.-Y.; Hua, T.; Tang, G.; Peng, X.-Q.; Hao, M.-H.; Zuo, Q.-T. Adsorption characteristics of cu(ii) and zn(ii) by nano-alumina material synthesized by the sol-gel method in batch mode. *Environ. Sci. Pollut. Res.* **2019**, *26*, 1595–1605. [CrossRef]
3. Mishra, S.; Bharagava, R.N.; More, N.; Yadav, A.; Zainith, S.; Mani, S.; Chowdhary, P. Heavy metal contamination: An alarming threat to environment and human health. In *Environmental Biotechnology: For Sustainable Future*; Sobti, R.C., Arora, N.K., Kothari, R., Eds.; Springer: Singapore, 2019; pp. 103–125.
4. Rehr, A.P.; Small, M.J.; Matthews, H.S.; Hendrickson, C.T. Economic sources and spatial distribution of airborne chromium risks in the u.S. *Environ. Sci. Technol.* **2010**, *44*, 2131–2137. [CrossRef]
5. Zhang, W.; Zhang, L.; Hua, T.; Li, Y.; Zhou, X.; Wang, W.; You, Z.; Wang, H.; Li, M. The mechanism for adsorption of cr(vi) ions by pe microplastics in ternary system of natural water environment. *Environ. Pollut.* **2020**, *257*, 113440. [CrossRef]
6. Cai, W.; Wei, J.; Li, Z.; Liu, Y.; Zhou, J.; Han, B. Preparation of amino-functionalized magnetic biochar with excellent adsorption performance for cr(vi) by a mild one-step hydrothermal method from peanut hull. *Colloids Surf. A Physicochem. Eng. Asp.* **2019**, *563*, 102–111. [CrossRef]
7. Yang, Z.-H.; Cao, J.; Chen, Y.-P.; Li, X.; Xiong, W.-P.; Zhou, Y.-Y.; Zhou, C.-Y.; Xu, R.; Zhang, Y.-R. Mn-doped zirconium metal-organic framework as an effective adsorbent for removal of tetracycline and cr(vi) from aqueous solution. *Microporous Mesoporous Mater.* **2019**, *277*, 277–285. [CrossRef]
8. Zhang, L.; Fu, F.; Tang, B. Adsorption and redox conversion behaviors of cr(vi) on goethite/carbon microspheres and akaganeite/carbon microspheres composites. *Chem. Eng. J.* **2019**, *356*, 151–160. [CrossRef]

9.   Javadian, H.; Ahmadi, M.; Ghiasvand, M.; Kahrizi, S.; Katal, R. Removal of cr(vi) by modified brown algae sargassum bevanom from aqueous solution and industrial wastewater. *J. Taiwan Inst. Chem. Eng.* **2013**, *44*, 977–989. [CrossRef]

10. Liu, W.; Jin, L.; Xu, J.; Liu, J.; Li, Y.; Zhou, P.; Wang, C.; Dahlgren, R.A.; Wang, X. Insight into ph dependent cr(vi) removal with magnetic fe3s4. *Chem. Eng. J.* **2019**, *359*, 564–571. [CrossRef]

11. Ren, L.; Dong, J.; Chi, Z.; Huang, H. Reduced graphene oxide-nano zero value iron (rgo-nzvi) micro-electrolysis accelerating cr(vi) removal in aquifer. *J. Environ. Sci.* **2018**, *73*, 96–106. [CrossRef]

12. Sharma, G.; Thakur, B.; Naushad, M.; Al-Muhtaseb, A.a.H.; Kumar, A.; Sillanpaa, M.; Mola, G.T. Fabrication and characterization of sodium dodecyl sulphate@ironsilicophosphate nanocomposite: Ion exchange properties and selectivity for binary metal ions. *Mater. Chem. Phys.* **2017**, *193*, 129–139. [CrossRef]

13. Zhao, R.; Li, X.; Sun, B.; Li, Y.; Li, Y.; Yang, R.; Wang, C. Branched polyethylenimine grafted electrospun polyacrylonitrile fiber membrane: A novel and effective adsorbent for cr(vi) remediation in wastewater. *J. Mater. Chem. A* **2017**, *5*, 1133–1144. [CrossRef]

14. Saleh, T.A.; Sarı, A.; Tuzen, M. Chitosan-modified vermiculite for as(iii) adsorption from aqueous solution: Equilibrium, thermodynamic and kinetic studies. *J. Mol. Liq.* **2016**, *219*, 937–945. [CrossRef]

15. Yang, R.; Li, H.; Huang, M.; Yang, H.; Li, A. A review on chitosan-based flocculants and their applications in water treatment. *Water Res.* **2016**, *95*, 59–89. [CrossRef]

16. Shen, H.; Durkin, D.P.; Aiello, A.; Diba, T.; Lafleur, J.; Zara, J.M.; Shen, Y.; Shuai, D. Photocatalytic graphitic carbon nitride-chitosan composites for pathogenic biofilm control under visible light irradiation. *J. Hazard. Mater.* **2021**, *408*, 124890. [CrossRef]

17. Pincus, L.N.; Petrović, P.V.; Gonzalez, I.S.; Stavitski, E.; Fishman, Z.S.; Rudel, H.E.; Anastas, P.T.; Zimmerman, J.B. Selective adsorption of arsenic over phosphate by transition metal cross-linked chitosan. *Chem. Eng. J.* **2021**, *412*, 128582. [CrossRef]

18. Wang, H.; Ding, F.; Ma, L.; Zhang, Y. Edible films from chitosan-gelatin: Physical properties and food packaging application. *Food Biosci.* **2021**, *40*, 100871. [CrossRef]

19. Bento, R.; Pagán, E.; Berdejo, D.; de Carvalho, R.J.; García-Embid, S.; Maggi, F.; Magnani, M.; de Souza, E.L.; García-Gonzalo, D.; Pagán, R. Chitosan nanoemulsions of cold-pressed orange essential oil to preserve fruit juices. *Int. J. Food Microbiol.* **2020**, *331*, 108786. [CrossRef]

20. Campos, L.M.; de Oliveira Lemos, A.S.; da Cruz, L.F.; de Freitas Araújo, M.G.; de Mello Botti, G.C.R.; Júnior, J.L.R.; Rocha, V.N.; Denadai, Â.M.L.; da Silva, T.P.; Tavares, G.D.; et al. Development and in vivo evaluation of chitosan-gel containing mitracarpus frigidus methanolic extract for vulvovaginal candidiasis treatment. *Biomed. Pharmacother.* **2020**, *130*, 110609. [CrossRef]

21. Bhardwaj, S.; Bhardwaj, N.K.; Negi, Y.S. Surface coating of chitosan of different degree of acetylation on non surface sized writing and printing grade paper. *Carbohydr. Polym.* **2021**, 117674. [CrossRef]

22. Srivastava, B.; Singh, H.; Khatri, M.; Singh, G.; Arya, S.K. Immobilization of keratinase on chitosan grafted-β-cyclodextrin for the improvement of the enzyme properties and application of free keratinase in the textile industry. *Int. J. Biol. Macromol.* **2020**, *165*, 1099–1110. [CrossRef]

23. Simonič, M.; Valh, J.V.; Vajnhandl, S.; Hribernik, S.; Kurečič, M.; Zemljič, L.F. Alternative cleaning of compost leachate using biopolymer chitosan. *Fibers Polym.* **2017**, *18*, 445–452. [CrossRef]

24. Zhang, W.; Lan, Y.; Ma, M.; Chai, S.; Zuo, Q.; Kim, K.-H.; Gao, Y. A novel chitosan–vanadium-titanium-magnetite composite as a superior adsorbent for organic dyes in wastewater. *Environ. Int.* **2020**, *142*, 105798. [CrossRef] [PubMed]

25. Kyzas, G.Z.; Bikiaris, D.N. Recent modifications of chitosan for adsorption applications: A critical and systematic review. *Marine Drugs* **2015**, *13*, 312–337. [CrossRef] [PubMed]

26. Doshi, B.; Ayati, A.; Tanhaei, B.; Repo, E.; Sillanpää, M. Partially carboxymethylated and partially cross-linked surface of chitosan versus the adsorptive removal of dyes and divalent metal ions. *Carbohydr. Polym.* **2018**, *197*, 586–597. [CrossRef]

27. Chen, Z.; Yao, X.; Liu, L.; Guan, J.; Liu, M.; Li, Z.; Yang, J.; Huang, S.; Wu, J.; Tian, F.; et al. Blood coagulation evaluation of n-alkylated chitosan. *Carbohydr. Polym.* **2017**, *173*, 259–268. [CrossRef]

28. Li, J.; Jiang, B.; Liu, Y.; Qiu, C.; Hu, J.; Qian, G.; Guo, W.; Ngo, H.H. Preparation and adsorption properties of magnetic chitosan composite adsorbent for cu2+ removal. *J. Clean. Prod.* **2017**, *158*, 51–58. [CrossRef]

29. Huang, Z.; Li, Z.; Zheng, L.; Zhou, L.; Chai, Z.; Wang, X.; Shi, W. Interaction mechanism of uranium(vi) with three-dimensional graphene oxide-chitosan composite: Insights from batch experiments, ir, xps, and exafs spectroscopy. *Chem. Eng. J.* **2017**, *328*, 1066–1074. [CrossRef]

30. Rasool, K.; Helal, M.; Ali, A.; Ren, C.E.; Gogotsi, Y.; Mahmoud, K.A. Antibacterial activity of ti3c2tx mxene. *ACS Nano* **2016**, *10*, 3674–3684. [CrossRef]

31. Naguib, M.; Mashtalir, O.; Carle, J.; Presser, V.; Lu, J.; Hultman, L.; Gogotsi, Y.; Barsoum, M.W. Two-dimensional transition metal carbides. *ACS Nano* **2012**, *6*, 1322–1331. [CrossRef] [PubMed]

32. Wang, J.; Ye, T.-N.; Gong, Y.; Wu, J.; Miao, N.; Tada, T.; Hosono, H. Discovery of hexagonal ternary phase ti2inb2 and its evolution to layered boride tib. *Nat. Commun.* **2019**, *10*, 2284. [CrossRef]

33. Chen, J.; Huang, Q.; Huang, H.; Mao, L.; Liu, M.; Zhang, X.; Wei, Y. Recent progress and advances in the environmental applications of mxene related materials. *Nanoscale* **2020**, *12*, 3574–3592. [CrossRef]

34. Zhang, Q.; Teng, J.; Zou, G.; Peng, Q.; Du, Q.; Jiao, T.; Xiang, J. Efficient phosphate sequestration for water purification by unique sandwich-like mxene/magnetic iron oxide nanocomposites. *Nanoscale* **2016**, *8*, 7085–7093. [CrossRef] [PubMed]

35. Shahzad, A.; Rasool, K.; Miran, W.; Nawaz, M.; Jang, J.; Mahmoud, K.A.; Lee, D.S. Two-dimensional ti3c2tx mxene nanosheets for efficient copper removal from water. *Acs Sustain. Chem. Eng.* **2017**, *5*, 11481–11488. [CrossRef]

36. Guo, J.; Peng, Q.; Fu, H.; Zou, G.; Zhang, Q. Heavy-metal adsorption behavior of two-dimensional alkalization-intercalated mxene by first-principles calculations. *J. Phys. Chem. C* **2015**, *119*, 20923–20930. [CrossRef]
37. Li, K.; Zou, G.; Jiao, T.; Xing, R.; Zhang, L.; Zhou, J.; Zhang, Q.; Peng, Q. Self-assembled mxene-based nanocomposites via layer-by-layer strategy for elevated adsorption capacities. *Colloids Surf. A Physicochem. Eng. Asp.* **2018**, *553*, 105–113. [CrossRef]
38. Fan, W.; Qiao, J.; Guan, X. Multi-wavelength spectrophotometric determination of cr(vi) in water with abts. *Chemosphere* **2017**, *171*, 460–467. [CrossRef]
39. Aljeboree, A.M.; Alshirifi, A.N.; Alkaim, A.F. Kinetics and equilibrium study for the adsorption of textile dyes on coconut shell activated carbon. *Arab. J. Chem.* **2017**, *10*, S3381–S3393. [CrossRef]
40. Chen, Y.; Zhang, D. Adsorption kinetics, isotherm and thermodynamics studies of flavones from vaccinium bracteatum thunb leaves on nka-2 resin. *Chem. Eng. J.* **2014**, *254*, 579–585. [CrossRef]
41. Zhou, C.; Wu, Q.; Lei, T.; Negulescu, I.I. Adsorption kinetic and equilibrium studies for methylene blue dye by partially hydrolyzed polyacrylamide/cellulose nanocrystal nanocomposite hydrogels. *Chem. Eng. J.* **2014**, *251*, 17–24. [CrossRef]
42. Liu, F.; Zhou, A.; Chen, J.; Jia, J.; Zhou, W.; Wang, L.; Hu, Q. Preparation of ti3c2 and ti2c mxenes by fluoride salts etching and methane adsorptive properties. *Appl. Surf. Sci.* **2017**, *416*, 781–789. [CrossRef]
43. Yu, Z.; Zhang, X.; Huang, Y. Magnetic chitosan–iron(iii) hydrogel as a fast and reusable adsorbent for chromium(vi) removal. *Ind. Eng. Chem. Res.* **2013**, *52*, 11956–11966. [CrossRef]
44. Geng, B.; Jin, Z.; Li, T.; Qi, X. Kinetics of hexavalent chromium removal from water by chitosan-fe0 nanoparticles. *Chemosphere* **2009**, *75*, 825–830. [CrossRef]
45. Lv, X.; Xu, J.; Jiang, G.; Xu, X. Removal of chromium(vi) from wastewater by nanoscale zero-valent iron particles supported on multiwalled carbon nanotubes. *Chemosphere* **2011**, *85*, 1204–1209. [CrossRef] [PubMed]
46. Shi, L.-n.; Zhang, X.; Chen, Z.-l. Removal of chromium (vi) from wastewater using bentonite-supported nanoscale zero-valent iron. *Water Res.* **2011**, *45*, 886–892. [CrossRef] [PubMed]
47. Lee, G.; Park, J.; Harvey, O.R. Reduction of chromium(vi) mediated by zero-valent magnesium under neutral ph conditions. *Water Res.* **2013**, *47*, 1136–1146. [CrossRef] [PubMed]
48. Ayyildiz, O.; Acar, E.; Ileri, B. Sonocatalytic reduction of hexavalent chromium by metallic magnesium particles. *Water Air Soil Pollut.* **2016**, *227*, 363. [CrossRef]
49. Siciliano, A. Removal of cr(vi) from water using a new reactive material: Magnesium oxide supported nanoscale zero-valent iron. *Materials* **2016**, *9*, 666. [CrossRef]

Review

Biotransformation of Citrus Waste-II: Bio-Sorbent Materials for Removal of Dyes, Heavy Metals and Toxic Chemicals from Polluted Water

Neelima Mahato [1,*], Pooja Agarwal [2,†], Debananda Mohapatra [1,†], Mukty Sinha [3,†], Archana Dhyani [4,†], Brajesh Pathak [5,†], Manwendra K. Tripathi [6,†] and Subramania Angaiah [7]

1   School of Chemical Engineering, Yeungnam University, Gyeongsan 38541, Gyeongsanbuk-do, Korea; debanandaiitb@gmail.com
2   Division of Chemistry, School of Basic and Applied Sciences, Galgotias University, Greater Noida 203 201, Uttar Pradesh, India; poojaocm5@gmail.com
3   Department of Medical Devices, National Institute of Pharmaceutical Education and Research, Ahmedabad, Palej, Gandhinagar 382 355, Gujrat, India; muktys@gmail.com
4   Department of Applied Sciences, School of Engineering, University of Petroleum and Energy Studies, Dehradun 248 007, Uttarakhand, India; adhyani@ddn.upes.ac.in
5   Department of Chemistry, Harishchandra PG College, Varanasi 221 005, Uttar Pradesh, India; brajeshpathak@gmail.com
6   Department of Metallurgical and Materials Engineering, National Institute of Technology, Raipur 492 010, Chhatisgarh, India; mktripathi.mme@nitrr.ac.in
7   Electro-Materials Research Laboratory, Centre for Nanoscience and Technology, Pondicherry University, Puducherry 605 014, Puducherry, India; a.subramania@gmail.com
*   Correspondence: neelapchem@gmail.com; Tel.: +82-010-2798-8476
†   Contributed equally as first author.

Citation: Mahato, N.; Agarwal, P.; Mohapatra, D.; Sinha, M.; Dhyani, A.; Pathak, B.; Tripathi, M.K.; Angaiah, S. Biotransformation of Citrus Waste-II: Bio-Sorbent Materials for Removal of Dyes, Heavy Metals and Toxic Chemicals from Polluted Water. *Processes* 2021, 9, 1544. https://doi.org/10.3390/pr9091544

Academic Editor: Jose Enrique Torres Vaamonde

Received: 3 August 2021
Accepted: 26 August 2021
Published: 30 August 2021

Publisher's Note: MDPI stays neutral with regard to jurisdictional claims in published maps and institutional affiliations.

Abstract: Industrial processes and anthropogenic activities generate huge amounts of wastes in the form of chemicals, such as heavy metals, dyes, fertilizers, pharmaceutically active chemicals, battery effluents and so on. When these chemicals are left untreated and discarded in the ground or surface waters, they not only cause pollution and harm the ecosystem but also cause toxic effects on the health of human beings, animals and food crops. There are several methods of removal of these toxic materials from the wastewaters, and adsorption by bio-sorbents has been demonstrated as one of the most inexpensive, efficient and convenient methods. Citrus is one of the largest grown fruit crops in the tropical and subtropical regions on the planet. After processing of the fruits at food processing industries, approximately half of the fruit mass is discarded as waste, which causes a number of pollution problems. Alternately, this biomass can be converted to bio-sorbents for the removal of harmful and toxic chemicals from the industrial effluents and wastewaters. The first part of this article contains a thorough review on the biotransformation of citrus waste for the production of biofuel and valuable compounds by fermentation involving microorganisms. The second and concluding part reviews the recent progress in biotransformation of citrus waste biomass (that may be remaining post-extraction of valuable compounds/biofuel generation) into efficient adsorbent substrate materials and their adsorption capacities. The article also includes the details of the synthesis process and mechanisms of adsorption processes.

Keywords: citrus waste biomass; bio-sorbent; pre-treatment; heavy metals; dyes; activated carbon; batch adsorption; fixed-bed adsorption column

1. Introduction

Citrus is one of the most popular and largest cultivated fruit crops in the tropical and subtropical regions on the planet, with an annual turnover exceeding 110–124 million tons. The fruits are largely processed in the food processing industries, and approximately half of the fruit mass (45–55%) is discarded as waste. The discarded mass consists of peels (flavedo and albedo), pith residue, seeds and parts inappropriate for human consumption. The waste, when directly discarded to the environment, causes huge problems in terms of pollution to the land and aquatic ecosystems, and underground as well as surface water resources. Alternately, this waste can be utilized as a sustainable and renewable natural resource and feedstock in a number of ways to obtain industrially important chemicals and valuable products by utilizing modern extraction methods, chemical processing techniques and biotransformation. On the one hand, utilization of the waste biomass provides an opportunity to produce valuable chemicals from green resources and avoid usage of harsh chemicals, and on the other hand, it helps in protecting the environment from the adverse effects of pollution. The authors have extensively reviewed the nature, properties and different technological processes of citrus waste valorization in their previously published articles [1–8].

1.1. Citrus Waste Pre-Treatment and Disposal

Citrus wastes produced by food processing industries are generally sent to waste-disposal plants, which require substantial transportation costs and availability of suitable sites for waste disposal. The most common waste management methods employed for the citrus wastes are composting, anaerobic digestion, incineration, thermolysis and gasification [9]. Commonly, in small processing units in the underdeveloped or developing countries, the wastewater is not believed to be toxic or harmful as sewage wastewater and is disposed of into drainage. When the dilute wastewaters, disposed of through drainage, reach streams, lakes, tidal waters, ponds or dumping wells, the organic solids carried by the wastewaters begin to decompose. During the process of decomposition, the dissolved oxygen present in the water is consumed. As a result, anaerobic or putrefaction reactions set in, and the aquatic organisms, such as insects, planktons (zooplanktons and phytoplanktons), fish, aquatic organisms, etc., die because of oxygen deficiency in the water [8]. Within the past three decades, there has been a multifold rise in the demand, and consequently, an increase in the production, supply, processing and applications of processed citrus products. At the same time, there has also been a huge rise in the quantities of dilute liquid and solid wastes, which are required to be disposed of appropriately. Large processing plants and units utilize large amounts of chlorinated waters to keep the processing tables, containers, vessels and equipment clean, and hence, the wastewater also contains chlorine. The effluent waters contain varied quantities of peel oil traces, pulp, juice sacs, follicles and organic materials [10]. Depending upon the concentration of materials and source of origin, the wastewater is divided into three categories, as follows:

(a) Dilute wastewaters: This includes, (i) fruit-rinsing water, (ii) surface condenser water and (iii) water from barometric condensers of evaporators. This disposal contains good quantities of carbohydrates and a low concentration of nitrogen, unlike domestic sewage which has the opposite composition, i.e., low in carbohydrates and rich in nitrogen. This can be released to water bodies without any fear of adverse effects to the ecosystem.

(b) Wastewater of intermediate concentration: This includes, (i) floor washing, (ii) equipment clean-up water and (iii) sectionizing wastewater. This contains solid waste concentrations ranging from trace amounts to ~2%. This requires some level of treatment prior to disposal to water bodies.

(c) Concentrated wastewaters: This includes, (i) dripping waters from can closing and filling machines, (ii) effluent from peel-oil centrifugal and (iii) waste alkali from sectionizing or evaporator cleaning. This contains 2–6% of soluble solids and high concentrations of organic materials.

When dilute wastewater is disposed into water bodies, lakes and basins located in far-off places from the residential areas, sandy beds or sandy lands, the liquid in the wastewater can either percolate down into the soil or evaporate easily, leaving behind the solids at the surface. Later, the clogged wastewaters were also suggested for spray irrigation with somewhat diluted concentration to pasture lands [11,12]. One such experiment conducted at pasture land bearing leguminous cover crops has been reported to show good results [13]. However, spray irrigation to the wood lands, pastures and green vegetation showed negative effects. For disposal, usually, several ponds are constructed, sometimes in series, to manage the waste to flow from one pond to another. The discharge into groves or wells may result in defoliation of trees, probably because of the loss of oxygen in the soil around the roots of the trees. Wastes released into the city sewage system may contaminate the underground water sources and cause damage to pumps and piping, clogging of sand beds and foaming in primary settling tanks. Additionally, pumping the wastewaters into the wells is prohibited because of probable contamination of underground water supplies. Furthermore, fermentation causes gaseous build-up inside the dumping wells and sometimes the pressure mounts to an excessive level to blow back and cause fire outbreaks. Some of the workable solutions in such cases are treatment with nitrogen, extensive aeration and chemical flocculation followed by lagooning [14–17]. Chemical flocculation with lime combined with aeration in order to promote floating of solids has been observed to result in a 64% reduction of suspended solids and up to a 30% reduction in biological oxygen demand (BOD) [18–20]. Although this treatment does not reduce the BOD of the wastewaters, it helps in fulfilling the purpose of waste pre-treatment by regulating the pH to a less acidic consistency and clarifying the suspended solid particles. Common methods of citrus waste removal from the food processing plants and industries, and its adverse effects on the environment, are shown in Figure 1.

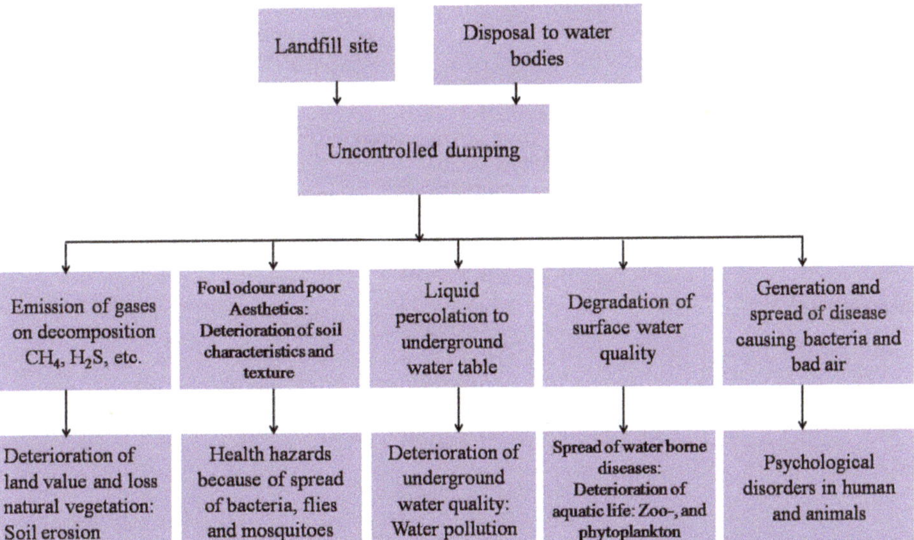

Figure 1. Conventional methods of waste removal from citrus processing plants and deteriorating effects to the environment.

The semi-solid citrus wastes have very high-water content, and they are difficult to dry through conventional methods. Furthermore, these processes consume huge energy. The most common method of solid and semi-solid biowaste management methods is composting. However, in case of citrus waste, it causes additional problems. Composting or digesting the citrus waste is not a practical choice as these contain large amounts of essential oils, mainly limonene, which inhibits microbial growth and the fermentation process and affects decomposition. Therefore, extraction of oils from the peel waste is very important before disposal to the landfills. Murdock and Allen reported that the oils present in the orange peels are toxic to yeasts [19]. These have been found to exhibit inhibitory effects on the growth of several useful bacteria, yeast and molds, e.g., *Bacillus subtilis,*

Saccharomyces cerevisiae and *Aspergillus awamori* [20]. Removal of oils from the citrus waste enables application of decomposition methods or anaerobic digestion, incineration, thermolysis and gasification. Detailed illustrations of practical methods of systematic and environment-friendly disposal of citrus wastes and their relative merits are displayed in Figure 2.

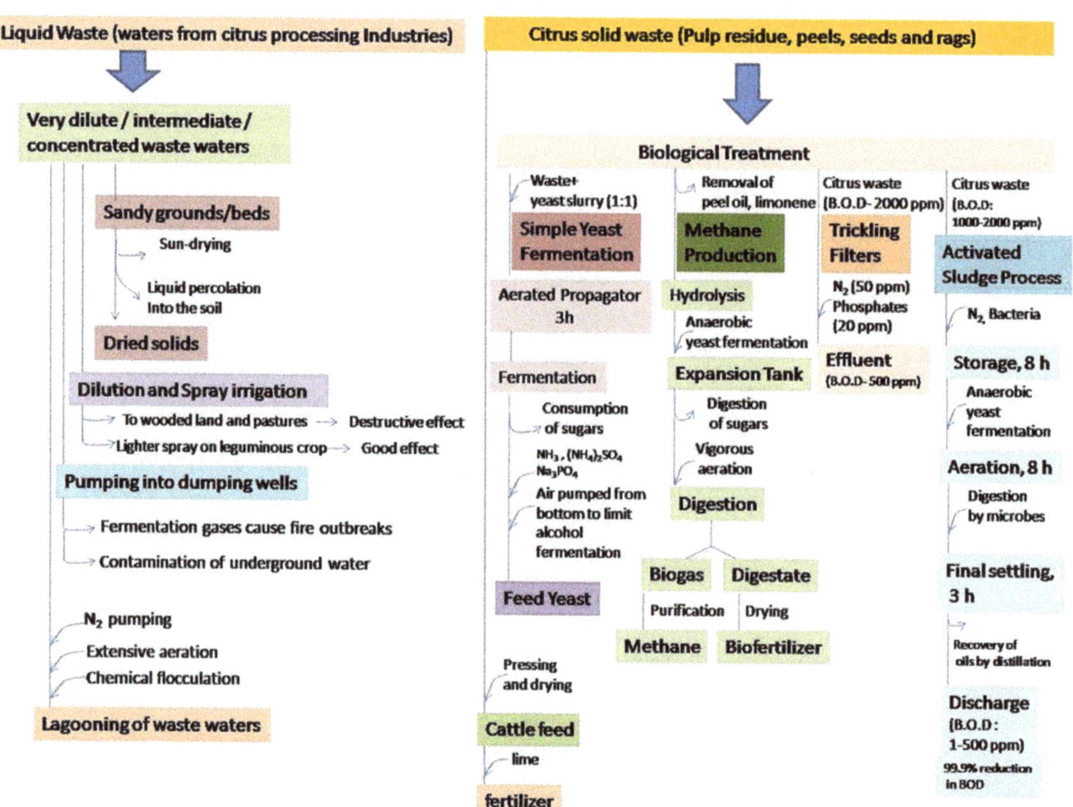

Figure 2. Methods of systematic and environment-friendly disposal of citrus wastes, conversion into useful materials and their relative merits.

Thus, it appears that the extraction of oil from the peel waste streams not only provides useful by-products, but also contributes toward pollution abatement. Citrus wastes also contain large amounts of sugars, which along with moisture, invite bacteria to grow. Decaying waste causes visual displeasure, odoriferous environment and attracts flies. Furthermore, citrus wastes are required to be processed quickly before compositional changes occur. Therefore, waste disposal to the dumping grounds has additional disadvantages. The nitrogen content in the solid waste materials (<0.14%) is quite insufficient to support bacterial decomposition. Therefore, to support decomposition, it robs oxygen from the soil underneath, resulting in a deficiency of nitrogen in the soil. The effect can be witnessed in terms of de-coloration of the vegetation or grasses at or surrounding areas of the dumping ground. The problem can be overcome by adding N_2 supplements in the form of chemical fertilizers. In this process, 200–400 pounds of calcium cyanide is added to each ton of ground waste, mixed thoroughly and allowed to dry until crumbly. Besides calcium cyanides, nitrates, ammonium sulfate and super phosphate are also sometimes added [10].

1.2. Pollutants: Dyes, Heavy Metals, Pharmaceutically Active Compounds (PACs) and Other Contaminants

Processes and manufacturing industries, such as metal plating, metal finishing, automotive, semiconductor manufacturing, pulp and paper production, mining operations, ceramics production, tanneries, radiator manufacturing, smelting and alloy manufacturing, battery manufacturing, corrosion of pipes and infrastructures, textiles and dye industries, etc., release a number of harmful and hazardous chemicals into wastewaters and effluents [21,22]. Researchers have reported that more than 700 kinds of pollutants, both organic and inorganic chemicals, mostly toxic and non-biodegradable in nature, are regularly discarded into the water bodies. The non-biodegradable pollutants are persistent in the environment. The heavy metals in the list of pollutants are cadmium, platinum, mercury, copper, lead, chromium, arsenic, antimony, etc., which have been observed to cause adverse health effects, such as gastrointestinal disorders, stomatitis, tumors, hemoglobinuria, ataxia, paralysis, diarrhea, neurological disorders, muscular dystrophy, vomiting, convulsions and so on [23–25]. Heavy metal pollution is a serious problem because of the metals' persistent nature, and their ability to enter and accumulate in the food chain [26–28].

Dyes are organic compounds used for coloring textile materials, paper, plastic, paints and synthetic coloring materials. There are about 40,000 dyes and pigments with approximately 7000 different kinds of chemical structures known to chemists. A dye substance has two parts: chromophores and auxochromes. The chromophores impart color whereas the auxochromes impart intensity for the dye. The dyes are classified as acid, base, reactive, direct, disperse, solvent, sulfur, vat, etc., and involve a wide variety of applications as well as application methods. A vast majority of dyes and pigments are non-biodegradable and persistent in nature. Textile and fiber industries employ approximately 10,000 different kinds of dyes for dying and printing of clothes and fabrics. The concentration of the dye bath during the dying processes ranges between 10 and 200 mg L^{-1}, which retains approximately 10–15% of its initial concentration post-dying and released into the effluent [29,30]. The dye concentration of approximately one ppm or less in the wastewaters has been considered as a potential threat to the environment and to human and aquatic lives [29]. Cyanides from industries released into the environment, particularly in wastewaters, have detrimental effects. Cyanides and their complexes have been demonstrated to have bioaccumulative properties that result in ecological deterioration [31]. Most of the industrial wastewaters contain F-CN, the simplest and most toxic form of cyanide, which is formed by dissociation of cyanide complexes during cyanide-based electroplating operations [32]. Chemical methods are usually employed for the conversion of free cyanide (F-CN) into a complex of NH^{4+} and $CHOO^{-}$, but most of the methods are expensive. Furthermore, they also produce harmful by-products which contribute to environmental contamination as well as being detrimental to the bioremediation processes [33]. Citrus wastes have been found to be useful for the conversion of F-CN into a complex of NH^{4+} and $CHOO^{-}$. Apart from F-CN, most of the wastewaters contain heavy metals and the presence of the metallic species (Ni, Zn, As, Cr, Hg and Cu), which have been observed to slow down the conversion of F-CN. The latter occurs as the metallic ions become attached to the hydroxyl groups of the absorbent material responsible for the removal. In addition, it has been reported that material obtained after acid hydrolysis of citrus solid wastes increases the catalytic conversion of F-CN by ~3.86-fold compared with the unhydrolyzed solid waste. The conversion has been found to increase linearly with an increase in pH and temperature [33]. Adsorption of heavy metal ions on the adsorbate material is fundamentally driven by electrostatic attraction between the oppositely charged ions. Both the heavy metal ions present in the wastewater and the functional groups (-COOH, -OH) present on the bio-sorbent material carry charges. The electrostatic attraction between the charges results in binding and ion exchange between Na^{+}, Mg^{2+}, K^{+}, Ca^{2+} and heavy metal ions (M^{2+}). However, the ion exchange predominates due to stronger binding forces between the charged species.

In recent years, pharmaceutically active compounds (PACs) are emerging as one of the most harmful contaminants in the natural and wastewater systems. PACs are the vast range of complex organic chemical formulations which possess a range of medicinal properties and are used in the treatment, control and eradication of diseases in both humans and animals. These include antibiotics, anti-inflammatories, cytotoxins, birth control pills, synthetic hormones and statins. The commonly encountered PACs in surface waters are erythromycin, metronidazole, sulfamethoxazole, trimethoprim, ciprafloxacin, amoxicillin, trimethoprim, tetracyclin, metformin, acetaminophen, diclofenac, ibuprofen, ketoprofen, naproxen, diazepam, fluoxetine and so on [34–37]. In the past few decades, consumption of pharmaceutical compounds has increased compared with many earlier decades in the century across the globe, resulting in a significant increase in the production of raw materials as well as the final PAC products. Manufacturing of PACs and life-saving drugs by the pharmaceutical companies and processing and packaging factories also releases huge amounts of compounds into the wastewaters. When excreted by humans or animals and disposed of inappropriately, these enter into the environment. In addition, PACs are also released by urban sewage, domestic hospital wastewaters, intensive livestock farming, liquid livestock manure production, sewage sludge from agricultural activities and effluents from sewage treatment plants [38]. The compounds undergo transformation into certain metabolites or breakdown compounds under the influence of temperature, light and vicinity of other chemical ingredients or microbes in nature under the process of biodegradation or photodegradation. Hence, the new molecules produced may sometimes be more toxic compared to their parent PACs. Furthermore, the transformation renders it difficult to monitor the presence of PACs in their original parent formulation in the environment (ground or surface waters). Presence of PACs in significant amounts in the drinking water has become an alarming concern in many countries. These compounds are capable of accumulating in the biological entities, primarily microbes, and making them drug-resistant. As a consequence, they lose their potency to act effectively against disease-causing microorganisms. Apart from drug resistance developments in disease-causing pathogens, PACs in the environment also cause feminization of male fish and amphibian species because of enhanced amounts of natural and synthetic estrogens in the habitat waters. In addition, increased amounts of PACs in irrigation waters to the crops produce food enriched with PACs. Such foods have been found to cause femaleness in human males, multiple organ complications/failure, genetic and hereditary diseases [38,39].

Diclofenac, a non-steroidal anti-inflammatory (NSAID) drug, has been extensively employed in veterinary usage, fever, pain and injury, livestock farming, fisheries and dairy industries, and has resulted in increased amounts of this chemical in the flesh of animals. When the animals die and are fed upon by predatory birds, diclofenac enters into their bodies and causes serious health consequences, leading to death and collapse of their population. Genus *Gyps.* is most severely affected and has been included in the category of global extinction risk. Some of the highly endangered species from genus *Gyps.* are oriental white-backed vulture (*Gyps bengalensis*), long-billed vulture (*Gyps indicus*) and slender-billed vulture (*Gyps tenuirostris*), and the decline in their population has been recorded to be more than 95% since the early 1990s. The decline has been observed to continue at an annual rate of 22% to 48% [40–43]. Diclofenac causes kidney damage, increases in the concentration of uric acid and crystal formation in serum and vital tissues, visceral gout, bone tumor and death of the predatory birds. Decline of predatory birds' populations in the ecosystem has been reported to cause a severe imbalance in the food chain and an alarming threat to the survival of the human population in Europe and Asia. Absence of the predatory bird population (eagles and vultures), or scavengers of wild and domestic ungulate carcasses in the ecosystem, has led to an increase in the population of feral dogs (*Canis familiaris*), and consequently, an increased risk of rabies spread in the human population. Furthermore, there has also been an alarming increase in the population of rats (*Ratus* spp.) and increasing risk of transmission of various diseases observed, such as bubonic plague, brucellosis, tuberculosis and anthrax in humans and livestock. Disappearing eagle and vulture populations have also created a huge challenge for the Parsee and Tibetese communities in India, Nepal and Tibet, as they find it difficult to continue their burial ritual/practices at sky burial sites and towers of silence [39].

1.3. Health Hazards of Pollutants

Metal industries related to mechanical works and battery manufacturing carrying out electroplating, metal plating, etc., release substantial amounts of heavy metals into the wastewaters. Additionally, large amounts of poisonous dyes are released from pigments or printing industries [44]. The pigment and dye industries produce different kinds of dyes, such as methylene blue (MB), oxazine and xanthenes compounds, azo dyes, methyl violet, etc. Among various dyes, methylene blue and heavy metal salts are the most frequently used chemicals for dying silk, wood and cotton [45]. MB is a common cationic dye extensively used in medical, textile and printing industries. The wastewater effluents from these industries contain high amounts of dyes, which can cause severe environmental pollution. It contaminates the water bodies, such as rivers, ponds, lakes, ditches and even the ground waters. These chemicals are toxic and cause severe impacts on human health as they may cause nausea, vomiting, diarrhea, etc., when ingested. Thus, removal of MB from wastewater is of great concern not only from an environmental point of view, but also for the sake of human life. Various methods used for dye removal include adsorption, flocculation, precipitation, ion exchange, electro-kinetic coagulation, ozonization and so on. Among the above-mentioned processes, adsorption is one of the most efficient methods due to its simple design, ease of operation and insensitivity to toxic substances [46]. The molecular structures of some of the popular anionic and cationic dyes are shown in Figure 3. The hazardous effects of dyes and heavy metals on human health, the ecosystem and the environment, and the possible use of citrus waste-derived bio-sorbents for the removal of harmful chemicals from the wastewaters, are illustrated in Figure 4. The permissible limits and the hazardous effects of heavy metals, dyes and other contaminants on human health have been summarized in Table 1.

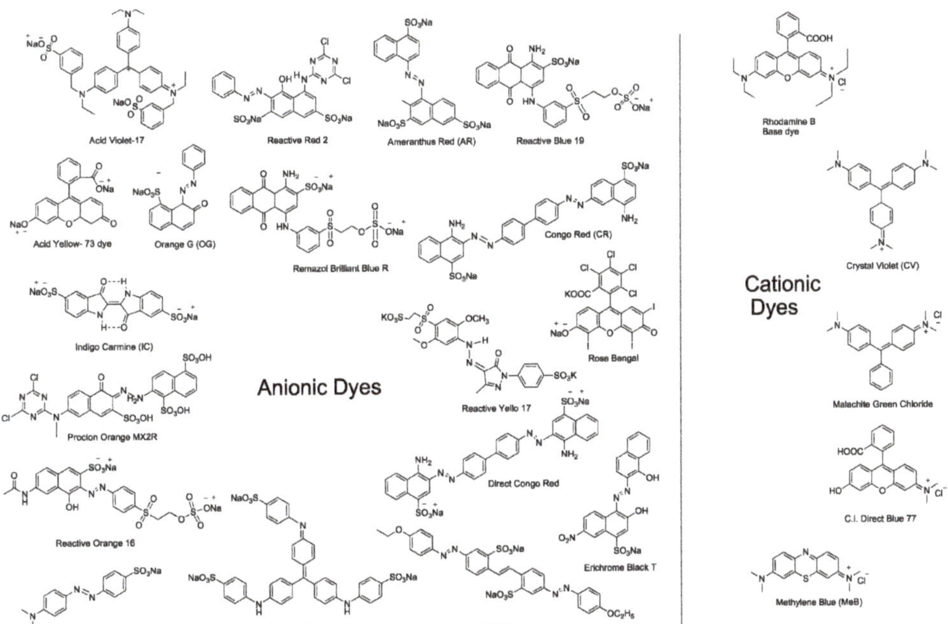

Figure 3. Structures of the anionic and cationic dyes.

Table 1. The permissible limits and the hazardous effects of heavy metals, dyes and other contaminants on human health.

Heavy Metals	Allowed Limits (WHO/EPA); mgL^{-1}	Source of Contamination	Adverse Effects on Health	Ref.
Arsenic	0.01	Erosion of natural deposits, runoff from soil in orchards, glass and electronics manufacturing waste, tanneries	Skin damage, deformation of digits, cancer, deterioration of circulatory system	[48]
Beryllium	0.004	Metal refineries, coal burning, discharge from electrical, aerospace and metal finishing factories/industries	Gastrointestinal disorders	[49]
Cadmium	0.005	Corrosion of pipes and infrastructures, erosion of metal in refineries, battery waste, paints and dyes	Kidney damage, cancer of lungs, malfunctioning of vital organs, proteinuria	[50]
Chromium	0.05–0.25	Steel industries, metal finishing factories, pulp mills and corrosion of stainless steel pipes and infrastructures, tanneries	Allergy, dermatitis, hemolysis, kidney failure, carcinomas, mutagenic diseases	[51]
Copper	1.0–1.3	Corrosion of household utensils and artefacts, plumbing system, erosion from copper mines	Gastrointestinal disorders, abdominal irritation, liver and kidney damage	[52]
Lead	0.005–0.015	Battery waste, corrosion of pipes and plumbing system, solder joints, erosion from natural deposits	Retardation of growth in children, abnormality in mental health and physical growth, anemia, vomiting, kidney and liver damage, high blood pressure	[49,53]
Mercury	0.002	Erosion from natural deposits, discharge from refineries and factories, runoff from landfills, croplands	Hypersensitivity, fever, vomiting, neurological disorders	[49]
Selenium	0.01–0.05	Petroleum refineries, erosion from natural deposits and mining sites	Loss of hair and finger nails, red skin, numbness in fingers and toes, burns, circulatory system disorders	[49]
Zinc	5	Electroplating industry, galvanization of metals, motor oil, battery waste, hydraulic fluid, tire dust	Corrosive to skin and eyes, zinc pox, sweet taste, throat dryness, cough, weakness, generalized aching, chills, fever, nausea, vomiting	[54]
Dyes	Less than 1.0 ppm	Effluents from fiber and textile industries, paper and pulp industries, plastic industries, paint industries	Nausea, vomiting, diarrhea	
Fluoride	1.5	Fluoride deposits (rocks such as topaz, cryolite and fluorapatite, etc.) on earth. Electroplating, glass, ceramics, steel manufacturing and phosphate fertilizer production; semiconductor manufacturing factories, pharmaceutical companies, beryllium extraction plants, aluminum smelters, fertilizer manufacturing and mining industries	Dental and skeletal fluorosis, crippling	[55,56]
PACs (Pharmaceutically active compounds)	$50\ ngL^{-1}$ to $0.1\ \mu gL^{-1}$	Pharmaceutical manufacturing plants, urban sewage, domestic hospital wastewaters, intensive livestock farming, liquid livestock manure production, sewage sludge from agricultural activities and effluents from sewage treatment plants	Genotoxicity and carcinogenicity, multiple drug resistance in pathogenic microorganisms, femaleness in males, tumor and vital organ failure in predatory birds	[49]

1.4. Citrus Peel-Derived Adsorbent Materials

The most popular adsorbent for the adsorption process is activated carbon. However, its use is still limited because it is expensive in terms of its high operational cost [57–59]. Advanced technology enables scientists to attempt to overcome the cost of the treatment process by using inexpensive, efficient and easily available adsorbents. In the literature, numerous studies have been reported to obtain low-cost activated carbons from agricultural wastes, such as wheat shells, rice husk, tea waste, neem leaf powder, cotton waste, banana peel and orange peel for the removal of heavy metals and other hazardous compounds from wastewaters [60–65]. However, the sorption potential of most of these low-cost sorbents is limited. The surface of the bio-sorbent can be modified to enhance its activity. Among many low-cost adsorbents, one such inexpensive and economical adsorbent precursor material is citrus fruit peel [66,67]. Citrus fruit peel has no use after the extraction of essential oils and other valuable chemicals, and thus, it is a commercially valuable and readily available resource for making adsorbent materials. Hence, conversion of citrus fruit peel to low-cost adsorbents serves multiple purposes. The unwanted waste can be converted to value-added products, such as low-cost adsorbent materials for the removal of heavy metals and dyes from aqueous solution. The latter efficiently alleviates the environmental pollution. Types of modification that can be introduced to the biomass-derived bio-sorbents are shown in Figure 5.

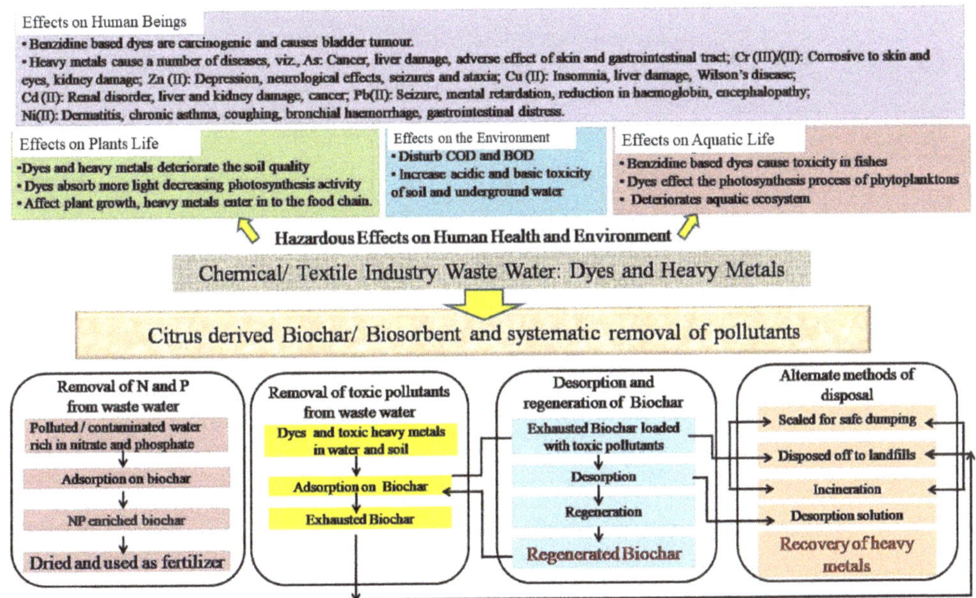

Figure 4. Hazardous effects of dyes and heavy metals released in the effluents from chemical and textile industries on humans and the environment, and the possible use of citrus waste-derived bio-sorbents for the removal of harmful chemicals from the wastewaters. Artwork developed from the information provided in [47].

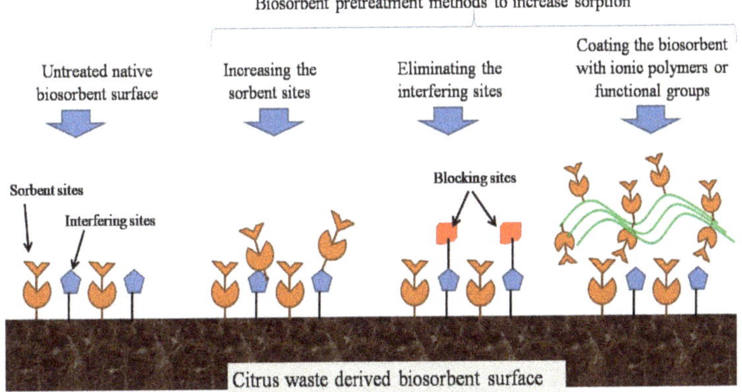

Figure 5. Basic principle of surface modification of citrus waste-derived bio-sorbent material by chemical pre-treatment for the removal and recovery of heavy/precious metals. Artwork created from the information provided in [68].

In this article, we focus on reviewing the recent progress in the field of citrus biomass research focused on producing energy, biofuels, important chemicals and bio-sorbent materials for the adsorption of dyes, heavy metals and toxic chemicals by processing of citrus waste biomass via biotransformation. Biotransformation is the modification of waste biomass or harmful chemical substances by certain microorganisms or chemical agents to render them either harmless to humans and the environment or for synthesis of useful products for safe consumption. The review has been presented in two parts under the titles *"Biotransformation of Citrus Waste-I: Production of Biofuel and Valuable Compounds by Fermentation"*, and *"Biotransformation of Citrus Waste-II: Bio-sorbent Materials for the Removal of Dyes, Heavy Metals and Toxic Chemicals from Polluted Water"*. The first part deals with the synthesis and production of biofuels (ethanol, methane and biodiesel) and valuable compounds, such as organic acids (citric, succinic, pyruvic, lactic, acetic), Vit-C, enzymes, single-cell proteins and prebiotics from fermentation of the citrus wastes [8]. In the second part of the article (present article), we attempt to review current and conventional trends of citrus waste disposal and their relative merits, and recent progress in biotransformation of citrus waste biomass into bio-sorbent material, employing physical, chemical or thermochemical methods for the adsorption of various pollutants, mainly heavy metals and dyes from polluted wastewaters and industrial effluents, and mechanisms and theoretical studies explored in the adsorption processes. The motivation behind this review is to conduct a detailed overview of the recent updates in this area of research and emphasize our focus towards the possibilities of harnessing the hidden potential of obtaining efficient products out of citrus biowaste, which otherwise is discarded in the dumping grounds as waste.

2. Methods of Preparation of Bio-Sorbents

Bio-sorbents from citrus wastes have been developed in a number of ways, such as (a) mechanical shredding/grinding, (b) physicochemical treatment, (c) thermochemical treatment and (d) biochemical methods using enzymes. Based upon the treatment, the different kinds of bio-sorbents obtained from citrus waste can be classified into the following categories: (i) native peel bio-sorbent, (ii) protonated peel bio-sorbent, (iii) peel pectic acid bio-sorbent, (iv) de-pectinated peel bio-sorbent, (v) carbonized peel-activated carbon bio-sorbent, (vi) chemically modified bio-sorbent and (vii) biochemically or enzymatically modified bio-sorbent. The native peel bio-sorbents are obtained from physical or mechanical treatment of citrus peel biowaste. The biomass is washed with tap water to remove dirt, followed by washing with distilled or deionized (DI) water, or nano-pure (NP) water, 3–5 times. Washing is a common step and is essentially included in all the techniques of biomaterial preparation. The native peel bio-sorbent is developed from drying the washed peel waste at 323 K, until it attains a constant weight, followed by grinding and sieving to a desired particle size of ~0.5–1.0 mm, appropriate for adsorption studies. Peel pectic acid is prepared from albedo of the citrus peel waste. The peels are treated with hot acidified water (HCl, pH 1.5) at 358 K and stirred for 2 h. The treatment at elevated temperature facilitates extraction of pectic acid from the peel tissues. The extracted pectic acid is then filtered and coagulated with an equal volume of ethanol (95%). The coagulate is washed with ethanol and dried until constant weight. The peel pectic acid is insoluble in aqueous medium and remains stable during the adsorption tests [40]. The de-pectinated peel adsorbents are obtained from the residue remaining after the extraction of pectin from peel waste. The pectin is mainly extracted from the albedo (white) part of the peels. The residue is washed thoroughly and dried in an oven until constant weight [1,2].

Chemical modification of bio-sorbents is carried out in order to enhance the adsorption capacity by introducing active functional groups by means of chemical reactions. Furthermore, it helps in enhancing the chemical stability and mechanical properties of the sorbent material. The latter prohibits the release of pollutants into the adsorption system. Chemical treatment is carried out via several methods. Some of the well-known methods are: (a) protonation (inorganic acid: HCl, H_2SO_4, HNO_3), (b) alkali saponification (using NaOH, $Ca(OH)_2$, $CaCl_2$), (c) phosphorylation, (d) blocking of functional groups using chemical reagents, (e) organic acids (citric acid, oxalic acid), (f) H_2O_2 treatment, etc. Protonated peel bio-sorbents are obtained from protonation of the citrus peel waste collected after washing and drying. The protonation is carried out by treating the material with acids, such as HCl or HNO_3. For this, the material is suspended in the acid and stirred or shaken for 6–8 h, followed by filtration, washing until neutral pH and drying until constant weight. Protonation is employed to remove excess of cations, such as Ca^{2+} or Na^+, present on the biomaterial surface, which interfere with the metal sorption process. These cations are replaced by protons, which enhances the binding of heavy metals by decreasing the competition between Na^+, K^+ and Ca^{2+} ions with heavy metals such as Cd^{2+}, Hg^{2+}, Pb^{2+}, Zn^{2+}, Se^{2+}, As^{2+} and so on. Replacement of Ca^{2+} and Na^+ ions by protons has exhibited enhancements in the adsorption of desired heavy metal ions and their removal from wastewater. Treatment with NaOH and citric acid is employed to introduce carboxyl groups on the bio-sorbent surface, which interact with heavy metals to form a complex, and the resultant structure helps in the removal of the toxic elements from the wastewater. Phosphorylation is employed to introduce abundant alcoholic "-OH" groups and phosphoric groups into the bio-sorbent material. The latter, possessing a high affinity for ferric iron, enhances the loading capacity for iron. Iqbal et al. carried out experiments to block "-COOH" and "-OH" functional groups using anhydrous CH_3OH, and concentrated HCl and HCHO, respectively. The modified bio-sorbents with blocked functional groups were found to exhibit a reduced adsorption capacity of Ni^{2+} by 78.57% and Zn^{2+} by 73.31%, confirming the main contribution of carboxyl and hydroxyl functional groups in the adsorption of heavy metal ions [69].

In the carbonization process, the dried peels are subjected to a very high temperature of ~773 K, followed by acid oxidation. An inert atmosphere, such as N_2, is employed to prohibit fire or rigorous oxidation. The positively charged amine groups present on the surface of the adsorbent material facilitate binding to anionic RMB reactive dye by electrostatic attraction [70]. Bhatnagar et al. prepared a bio-sorbent from lemon peels by thermal activation at 323 K in the presence of air, which converted the peels to ash. Treating bio-sorbents with H_2O_2 has been employed to avoid the release of color in bio-sorbents [71]. Treating with 1% NaOH and ethanol removes lignin and colored pigments. Carbonization through the chemical activation method is one of the most favored methods for preparation of adsorbent material from citrus fruit peel. Weight ratios of peel vs. activating agent, temperature and time of carbonization are the selected parameters for optimizing the preparation of an efficient adsorbent material [72]. Generally, the dried citrus fruit peel is fed to a mixer grinder, and the ground powder is mixed with activating agents such as ortho-phosphoric acid, zinc chloride or sulfuric acid. This is then carbonized in a muffle furnace by heating it at an elevated temperature of ~723–823 K, up to a duration of 0.75 to 1.5 h. The weight ratio of dried citrus fruit peel to activating agent varies in the range of 1:1 to 3:1. The charred material is then cooled and washed with dilute ammonia solution and distilled water. This removes any unconverted activating agent from the carbonaceous material. The washing of the sample is continued until the pH becomes neutral. The charred material is then left for drying overnight under ambient conditions. The dried samples are then crushed and fractioned into different sizes [72].

Cross-linked hydrogel adsorbents for the removal of dye molecules can be prepared by treating the ground peel powder with N-vinyl-2 pyrrolidone (NVP), followed by irradiation with gamma rays. Mahmoud et al. demonstrated that a gamma irradiation dose of 30 kGy to the hydrogel precursor composition of NVP and orange peels in the ratio of 1:1 results in optional homogeneity of the bio-sorbent material with appropriate properties for practical applications. Usage of hydrogel (cross-linked polymerized hydrogel) enables the material to adsorb and retain large volumes of water and facilitate in increasing the contact time between the pollutant dyes and adsorbent material by the virtue of its cross-linked three-dimensional network structure. A porous material possesses additional benefits in terms of extended surface area for adsorption [73]. Chemical treatment by formaldehyde and urea carried out by Rabia et al. demonstrated an enhancement in roughness and unevenness, with apparent pores and canals of irregular shapes in a 3D network structure of the material surface to facilitate extended contact time between the adsorbate molecules and adsorbent material, and enhance physisorption [74]. Treating with HCHO improves the shelf-life of the bio-sorbent and prevents microbial damage [75]. While the functional groups, such as carboxylic acid, amine and sulfonic acid groups participate in electrostatic interaction and binding with dye molecules, the bulky structure of the hydrogel can be conveniently collected, separated and regenerated either by washing with water or treating with acid for further usage. Dev et al. reported on the adsorption of selenium from wastewater by citrus peel-based bio-sorbents chemically modified by calcium alginate. The chemical modification provides structural stability to the adsorbent material, which allows its reusability. Besides, it enhances the number of "-COOH" and "-OH" functional groups due to alginate beads, which enables the bio-sorbent material to absorb and remove other metallic ions in addition to SeO_4^{2-} and SeO_3^{2-} [76].

Kam and Lee carried out adsorption of amoxicillin onto the activated carbon surface prepared from citrus peel waste from aqueous solution containing the antibiotic, and reported an efficient adsorption within 30 min and attainment of equilibrium in 90 min. The waste citrus peel-based activated carbon showed a maximum adsorption capacity of 125 mg/g of the adsorbent at 293 K [77]. Putra et al. studied adsorption and removal of amoxicillin using commercial activated carbon [78]. Moussavi et al. reported on the adsorption studies of amoxicillin on commercial activated carbon and activated carbon derived from pomegranate wood [79]. Ding et al. developed activated carbon from sewage sludge and oil sludge and reported adsorption and removal of oxytetracycline and chlortetracycline [80]. Baccar et al. obtained activated carbon from olive-waste cake to absorb naproxen, ketoprofen, diclofenac and ibuprofen from aqueous solution containing the contaminants [81]. Ahmed et al. produced activated carbon from Siris seed pods and carried out adsorption and removal of metronidazole from contaminated water [82]. On the other hand, activated carbon developed from vine wood by Pouretedal and Sedech et al. showed efficient adsorption of amoxicillin, cephalexin, tetracycline and penicillin from contaminated water [83].

The adsorption of heavy metal ions onto the adsorbent surface is influenced by several factors, such as nature of the material, charge on the chelating metal ion, size of the ion, nature of donor atom present in the ligand, buffering environment during the adsorption process and exchange of ions, nature and properties of the solid support, and so on [84]. In this direction, Li et al. carried out chemical modification employing 20% isopropyl alcohol to remove coloring compounds from orange peels, along with polar compounds. This was followed by saponification by addition of 0.1 M NaOH/0.1 M NH_4OH/saturated solution of $Ca(OH)_2$. The Na^+/NH_4^+ or Ca^{2+} ions become attached to the cellulose molecules of the adsorbent material and facilitate an ion exchange mechanism between Na^+, NH_4^+ or Ca^{2+} ions and the bivalent heavy metal ions. In the next step, the saponified orange peels are treated with 0.6 M acid at an elevated temperature of 353 K. The heat is required to produce a condensation product and acid anhydride. The latter combines with cellulose hydroxyl groups and results in the formation of ester linkage and introduction of carbonyl groups to the cellulose molecule. The additional carbonyl functional groups introduced to the cellulose molecule enhance metal ion adsorption [85]. In another method, Liang et al. treated orange peels with NaOH followed by mercapto-acetic acid ($C_2H_4O_2S$) in order to convert the hydroxyl groups present in the cellulose molecules into mercapto groups. The latter exhibited a higher affinity towards heavy metal adsorption (Cu^{2+} and Cd^{2+}) from aqueous solutions [86]. The final different sized samples are used for adsorption purposes. The different methods of bio-sorbent pre-treatment by heat, chemical(s) and enzymes are summarized in Figure 6.

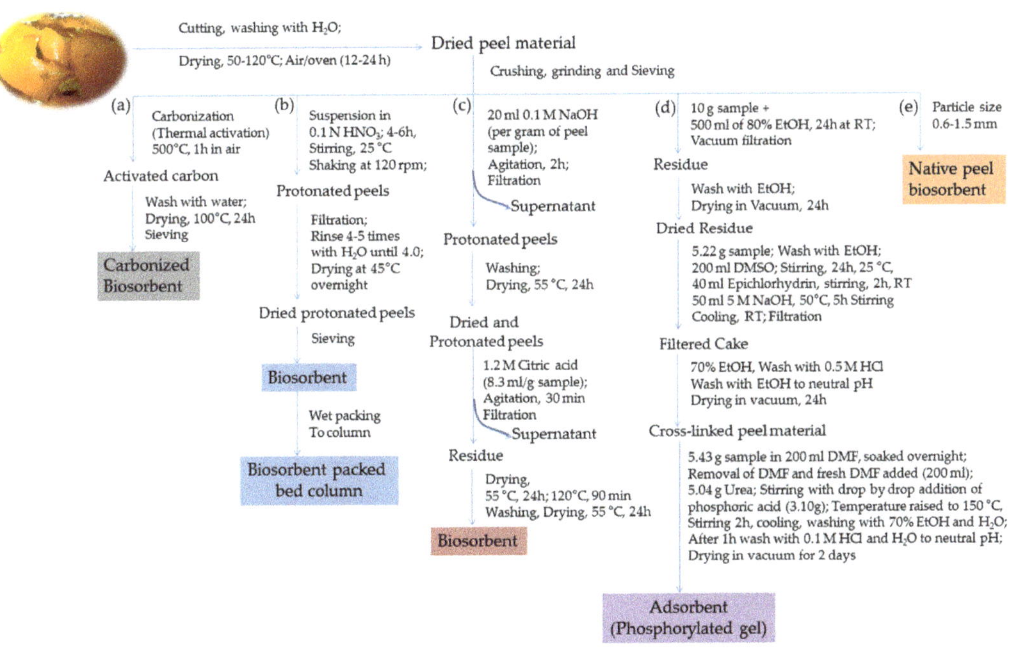

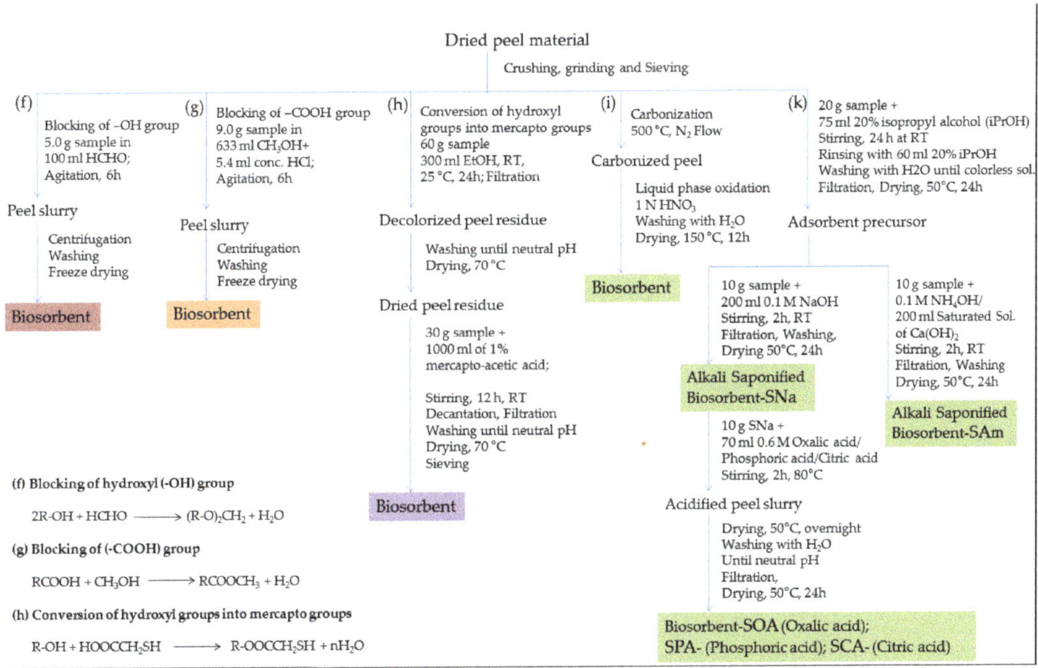

Figure 6. *Cont.*

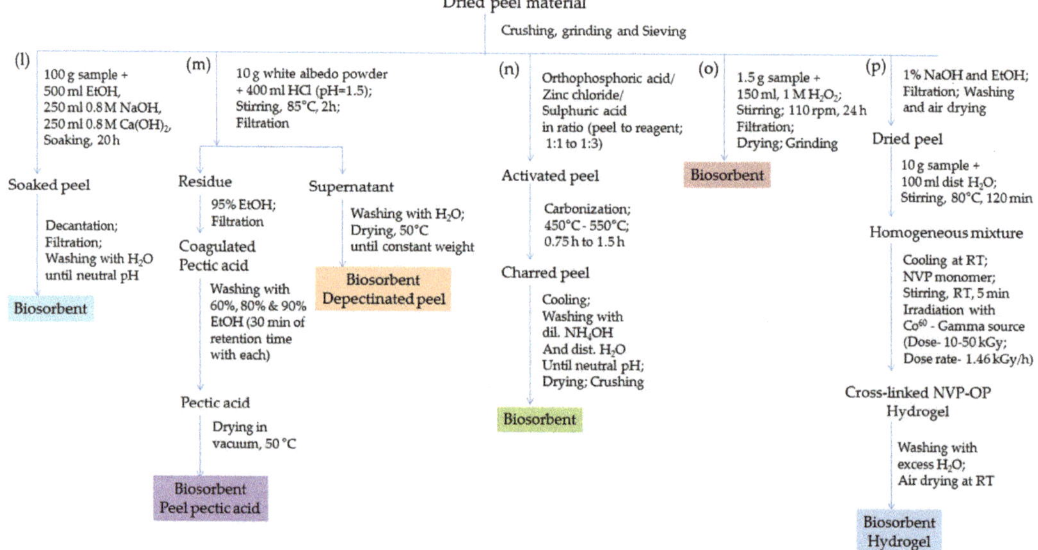

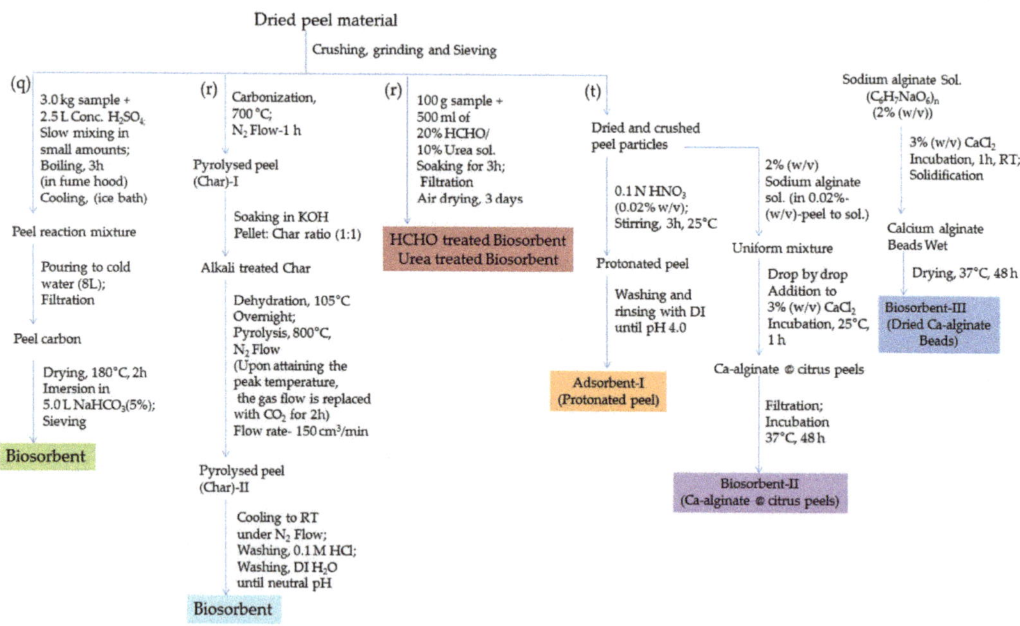

Figure 6. (a–t) Different methods of pre-treatment of the precursor material for the preparation of bio-sorbents from citrus wastes [69–76,84–88].

Cameron et al. carried out an elaborated study on the adsorption of Pb^{2+} ions by adsorbents synthesized from citrus peels and peel-derived pectin and concluded that fragmentation of larger molecules into smaller fragments and demethylation of the same occur. The latter plays an important role in the enhancement of the sorption capacity of pectin and derived materials. The fragmentation can be carried out either chemically or enzymatically. Pectin, a polysaccharide present in the citrus peels, can be modified via enzymatic or chemical conversion to develop a suitable bio-sorbent as well as fine-tune the desired properties, e.g., ion exchange and adsorption properties. Pectin polymer is made of galacturonic acid (GA) monomer units, a major sugar found in citrus fruits. It is basically concentrated in the linear homogalacturonan region (HG), which is pectin's dominant structural domain. The carboxylic acid functional groups present in the GA molecular structure interact with the heavy metal cation present in the wastewater or industrial effluents and require removal. While the polyanionic character of pectin is crucial for adsorption, masking of the negative charge present on the carbonyl group in the GA molecule by means of methyl esterification at C-6 position hence alters the overall functionality of the pectin molecule. In other words, the pectin functionality is dependent upon the total amount of methylation of GA units or degree of methylation (DM) and distribution of methylated GAs and non-methylated GAs (GAs with unmasked or free carboxylic acid functional groups) in the pectin polymeric chain in the HG region [89]. The de-esterification of the GAs or de-methylation of GAs can induce ordered or random distribution of de-methylated GAs in the pectin polymer chain. Both kinds of specific properties obtained post-modification, e.g., degree of methylation (DM) and degree of polymerization (DP), in pectin molecule have been reported to exhibit effects in terms of interaction with cation and sorption properties [87,88]. Pectin extraction can be carried out via aqueous extraction using an aqueous acid or base followed by purification of liquid extracts containing hydrocolloids, and isolating the extracted pectin from the mixture [1,2]. There is a basic difference in the product quality of pectin obtained post-acid/alkali treatment. The pectin obtained from the alkali extraction process contains a low degree of esterification (low DE pectin). The latter results from saponification of the ester groups present in the polymer molecule by alkali. On the other hand, pectin obtained from the acid extraction process contains a high degree of esterification (with high DE pectin ~50% and greater) [89]. The methods of obtaining bio-sorbent materials from citrus peel-derived oligosaccharides and enzymatically modified pectin have been summarized in Figure 7a, and their respective adsorption capacities towards Pb^{2+} are shown in Figure 7b.

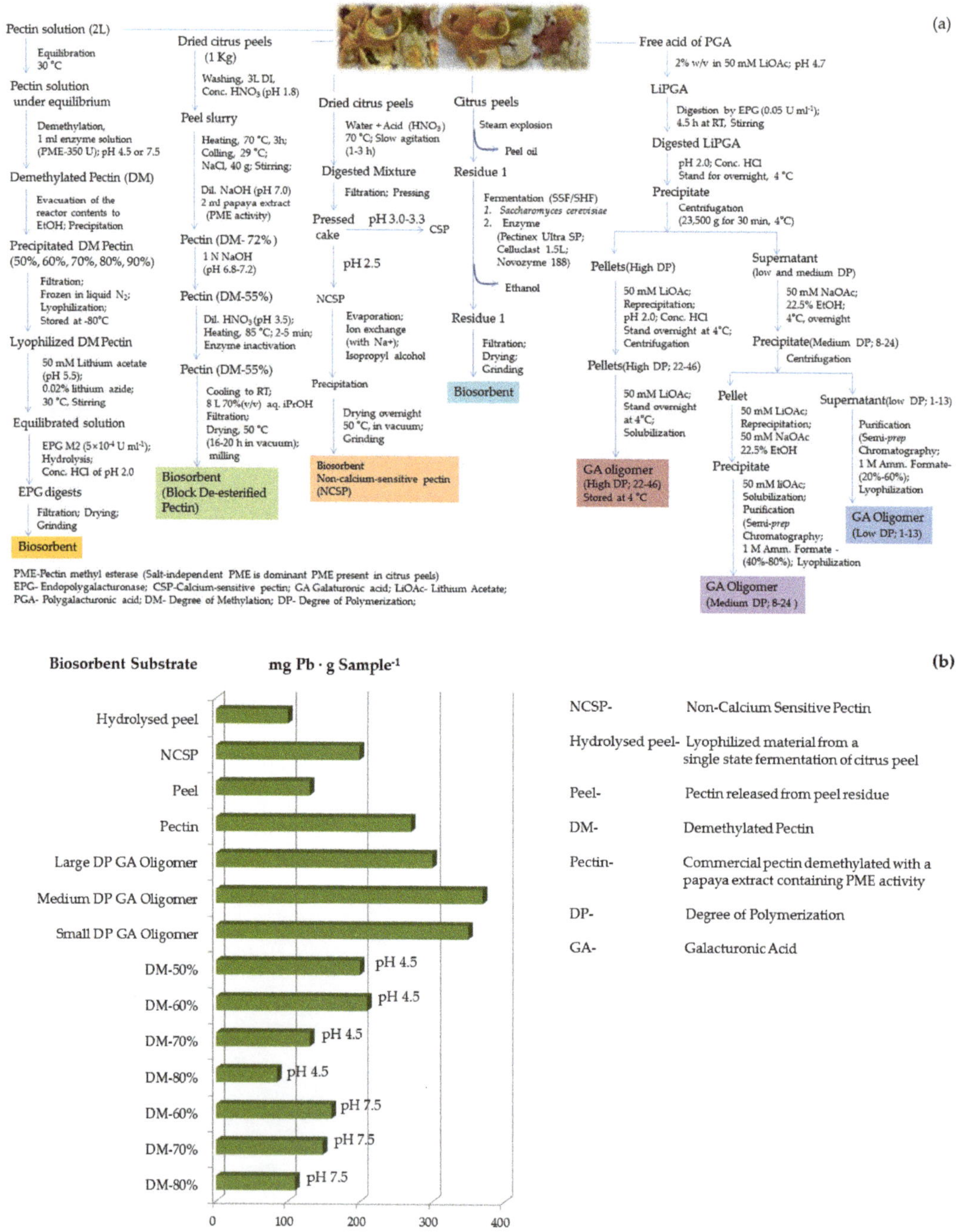

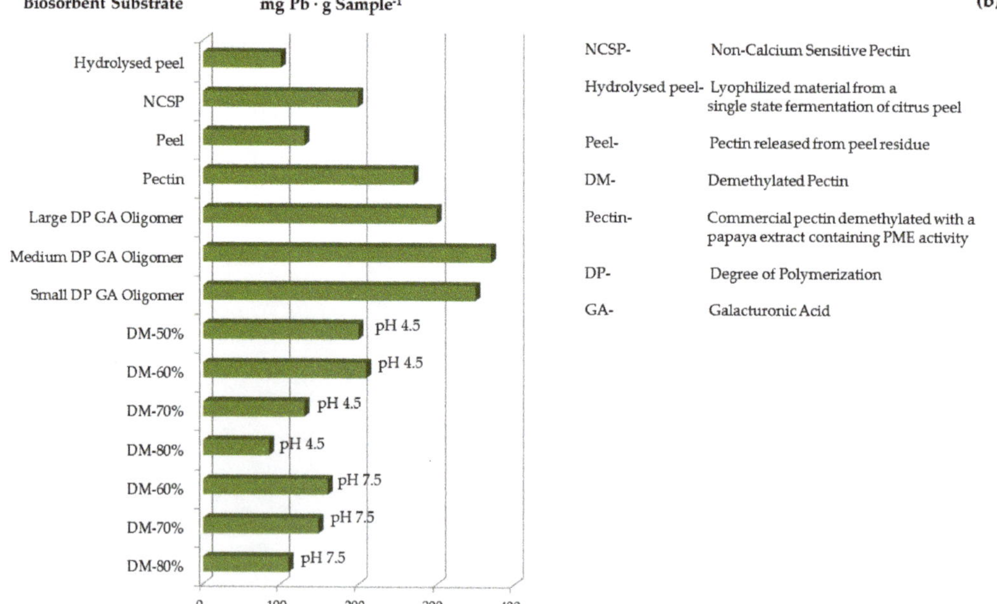

Figure 7. (**a**) Preparation of bio-sorbent from biochemical treatment of citrus peel waste. (**b**) Bio-sorption capacities of different bio-sorbent substrates prepared in (**a**). Adapted from the information provided in References [89–91].

3. Adsorption Experiments and Mechanism

The bio-sorption of harmful chemicals from wastewaters has been focused on achieving two prime targets: (a) development of novel bio-sorbent substrate material, and optimization of the adsorption process in terms of uptake of pollutant molecules/ions in a batch reactor containing a single target metal, and (b) enhancement of sorption uptake capacity by suitable processing [92,93]. From the viewpoint of carrying out adsorption processes/experiments, either or both of the two main techniques are employed, namely, batch adsorption test and/or column adsorption test. In the batch adsorption process, ion exchange has been identified as the main mechanism for the adsorption and removal of heavy metal ions or dye molecules. The carboxyl and hydroxyl functional groups present on the adsorbent substrate have been demonstrated to play key roles in the adsorption process. Other functional groups, such as amide, sulfonate, phosphate and amino groups, have also been reported to participate in the adsorption process. The adsorption process is a complex interplay between a number of mechanisms, such as complexation, coordination, chelation, ion adsorption or exchange, micro-precipitation, electrostatic interaction, H-bonding and so on. In the experimental part, a fixed amount of bio-sorbent substrate/material is placed with a definite volume of solution containing the pollutant or toxic metal ions/dyes, and stirred/shaken for a specified duration in a beaker or flask. After the adsorption process is completed, the solution is filtered and the adsorbent is regenerated or recycled by washing with water or treating with chemicals, such as acids, alkali or organic solvent(s). Alternately, the bio-sorbent material can also be regenerated by physical or thermo-physical treatments, such as heating, microwaving or sonication [93–95].

In some cases, the test solution has been observed to develop a brown color, which is explained by leaching out of carotenoids from the citrus peel biomass. Carotenoids are responsible for binding the heavy metal ions present in the test solution and forming complexes. The latter has been found to be unable to adsorb onto the bio-sorbent surface, and hence, left in the test solution imparting color. This problem can be solved by chemical pre-treatment of the bio-sorbent material, which helps in leaching out the carotenoid or other colored materials, such as chlorophyll, from the citrus peel biomass. Treating the biomass with isopropanol helps in the removal of soluble compounds without any adverse effect on the nature of biomass material or ion-binding sites on the surface. Citric acid treatment helps in the dissolution of polysaccharides present in the cell wall of the biomass. This facilitates in opening up the physical structure of the biomass and thereby increases the number of adsorption/binding sites, i.e., functional groups. Alkali treatment has been observed to impart a stronger effect on cell wall rupturing and facilitate the exposure of functional groups. Furthermore, the hydroxyl or carboxylic acid groups are converted to their salt forms, thereby helping to enhance the adsorption process by an ion exchange mechanism. An increase in temperature during alkali or acid pre-treatment has been observed in biomass loss. The factors influencing the bio-sorption process are initial pH of the test solution, concentration of the test solution (heavy metal ions, dye molecules, etc.), dosage of adsorbent, pre-treatment of the bio-sorbent, temperature during the adsorption process and duration of the contact time between the sorbent material and the test solution. The advantages of the batch adsorption technique are short analysis time, low operational costs, simple maintenance and conductance and that it can be operated with locally available adsorbent materials with satisfactory efficiency [96].

A next-level technique, which can be employed for large sample sizes or industrial scale purposes regarding adsorption and removal of pollutants, is fixed-bed reactors or the column adsorption method. It is also called a continuous flow column system. The fixed-bed column is a commercially viable technique, and at present, employed in industries fitted with ion-exchange resins or other commercially available adsorbent materials. The column material is usually made of clear-extruded acrylic. The length of the column ranges between 25 and 30 cm with an internal diameter of 1.3–1.5 cm. The bottom layer of the column is filled with spherical glass beads (of diameter 3 mm) to facilitate even distribution of influent flowing across the column length. A fiber screen is placed at the top of the column to prohibit the scattering of the adsorbent material in the fixed-bed and confine it appropriately in a fixed position during the entire operation. The wet-packing of the column is usually performed in such a way so as to allow a void space inside the column up to 70%. The bed-height is hence adjusted up to 24 cm in order to enhance the length of the fixed-bed, and the column length can be increased by supplementing an additional number of columns in series. During operation, a certain amount of material by weight is wet-packed in the column and the column is conditioned with a suitable solvent (e.g., acidified water of pH 4.0–5.5) overnight. Post-conditioning, the test solution (prepared at the same pH) is percolated through the column at a constant flow rate using a peristaltic pump or a micro-tube pump. The test solution passes through the column contents, i.e., packed adsorption bed, and the latter adsorbs the heavy metal ions or dye molecules, leaving behind the solution with a lesser or minimal concentration of pollutants. If the driving force (concentration variation between sorbent and influent) remains high, the sorbent shows high metal ion uptake. This is due to the adsorbent material being saturated at a relatively high influent concentration, whereas the progressing metal solution comes repeatedly in contact with a fresher and more efficiently adsorbing surface, and thus the effluent leaving the column becomes virtually free of metal ions. The saturated column bed can be regenerated and recycled. Regeneration of the column bed is carried out by eluting the bed with a suitable desorption solution. The effluent samples are collected at a fixed interval of time using a fraction collector [97,98]. The schematic representation of the experimental set-up for the fixed-bed adsorption column is shown in Figure 8. The important reported results of experiments on bio-sorbent materials developed from citrus peel waste are listed in Tables 2–4. The various mechanisms of adsorption of heavy metals and dyes on the biomass derived bio-sorbent surface are shown in Figure 9.

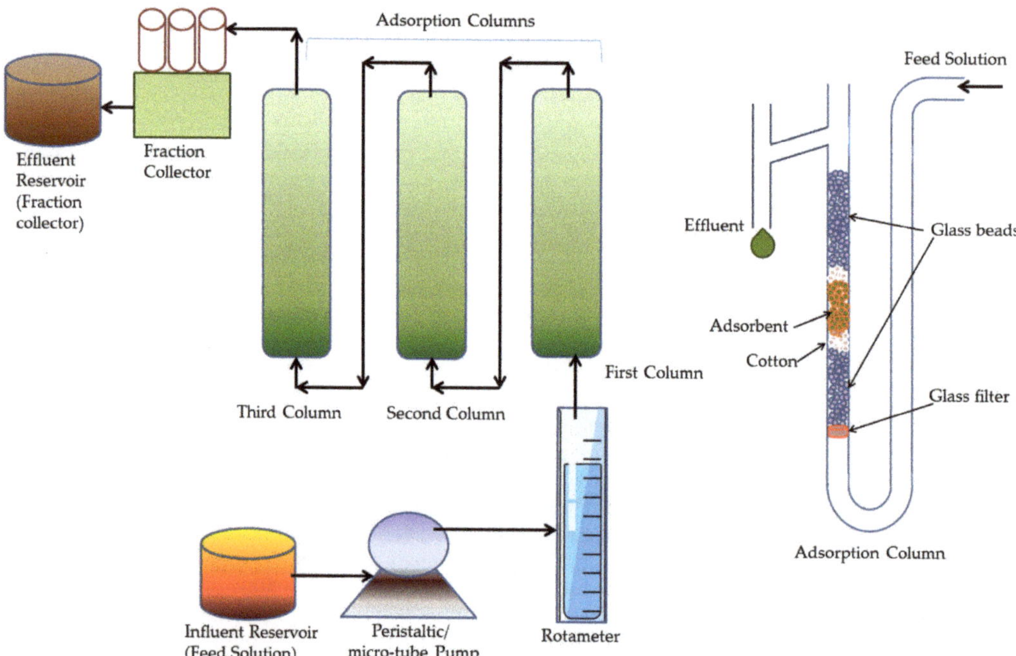

Figure 8. Schematic representation of experimental set-up for fixed-bed adsorption column. Artwork developed from References [97–99].

Table 2. Citrus waste-based bio-sorbents for removal of heavy metals from wastewater (BAT: Batch Adsorption Test; DAT: Direct Adsorption Test (Column Adsorption)).

Adsorbent	Heavy Metals	Processing Method	Maximum Adsorption Capacity (Per Gram Adsorbent)	Ref.
Citrus limetta peels	Uranium	Wash; Dry—333.15 K, 24 h; Grind, P.S. = 850 µm–1.2 mm; BAT—0.1 g BS in 40 mL MS; [U(VI)] = 250 mg/L; pH 2.0–7.0; CT—3 h; SRs—150 rpm	Maximum adsorption of 75.33 mg/g of bio-sorbent material at pH 4.0. Adsorbed species detected are UO_2^{2+}, UO_2OH^+, $(UO_2)_3(OH)^{5+}$, $(UO_2)_2(OH)_2^{2+}$	[100]
Citrus paradisi (Grapefruit) peels	Arsenic	Wash; Dry—333.15 K, 6 h; Grind, P.S.=100–125 µm; Wash; Dry—333 K, 24 h BAT—0.2 g BS in 15–20 mL MS; [As(V)] = 0.1–50 mg/L; pH 4.0–7.0; CT—15 min–2 h; SRs—100–200 rpm	Maximum adsorption of 37.76 mg/g of bio-sorbent material at pH 4.0, 318 K. 76–94% removal in different polluted water sources	[101, 102]
Orange peels	Nickel	Wash; Dry—373 K, 12 h; Grind, P.S. = 1.80 mm BAT—0.2 g BS in 100 mL MS; $[Ni^{2+}]$ = 10–200 mg/L; pH 5.0; SRs—200 rpm	1.05–29.04 mg/g adsorbent from metal solution of concentration 10–200 mg/L. Maximum adsorbing capacity of 33.14%	[103]
Orange peels	Zinc	Wash; Dry—333.15 K, CW; Grind, P.S. = 0.15–1.5 mm BAT—0.2 g in 50 mL MS, $[Zn^{2+}]$ = 100 mg/L; pH 4.0–6.0; CT—3 h, stirring DAT—Acrylic column, 50 cm length; 2.2 cm diameter; 24 g peel pack, $[Zn^{2+}]$ = 20, 30, 40 mg/L, pH 4.0; FL—8.5 mL/min	BAT—0.664 mmol/g (75%) at pH 6.0 DAT—0.42–0.44 mmol/g	[104]
Citrus reticulata (Ponkan mandarin) peels	Nickel, Cobalt, Copper	Wash; Dry—333.15 K, 24 h; Grind, P.S. $\leq$ 0.6 mm BAT—0.1 g BS in 25 mL MS, $[M^{2+}]$ = 0.01 M, pH = 4.8, Stirring—2 h DAT—50 cm length, 0.5 cm diameter, 1.0 g BS, $[M^{2+}]$ = 5 × 10^{-4} M, FS—3.5 mL/min, pH 4.8	BAT—Nickel—1.92 mmol/g; Cobalt—1.37 mmol/g; Copper—1.31 mmol/g DAT—Nickel—1.85 mmol/g; Cobalt—1.35 mmol/g; Copper—1.30 mmol/g	[105]
Citrus aurantium (Bitter Orange) fruit parts	Cobalt	Wash; Sundry; Grind, P.S.= 250 µm BAT—2 g BS in 50 mL MS; $[Co^{3+}]$ = 5 mg/L; pH = 2.0, CT = 90 min; stirring	Flavedo = 57.99%; Albedo = 20.11%; Juice = 15.63%; Segment membrane = 20.90%; Seeds = 10.06%	[106]
Citrus reticulata Tangerine	Lanhanum Cerium	Wash; Dry 75–353 K, 24 h; Grind, P.S.= 355 µm BAT—0.5–3 g/l BS; $[M^{3+}]$ = 10–200 mg/L; pH 2.0–6.0; CT = 5–150 min; SRs—200 rpm; T 20–323 K	La(III)—154.86 mg/g Ce(III)—162.79 mg/g	[107]
Ponkan mandarin peels	Lead	Wash; Sun dry—7 d; Grind, P.S. $\leq$ 600 µm BAT—0.2 g BS in 25 mL MS; $[M^{2+}]$ = 0.5–1 g/L; pH 5.0; CT = 120 min; SRs—120 rpm; T 298 K	112.1 mg/g	[108]
Orange waste	Binary systems Cd^{2+}–Zn^{2+} Cd^{2+}–Pb^{2+}, Pb^{2+}–Zn^{2+}	Wash; Dry; Grind, P.S.= 0.6–1.5 mm; BAT—0.4 g BS in 100 mL MS; $[M_1^{2+}$–$M_2^{2+}]$ = 15–100 mg/L —15–25 mg/L added subsequently in 30 combinations.	Pb^{2+} > Zn^{2+} > Cd^{2+} Maximum uptake of 0.25 mmol/g adsorbent	[109]
Orange Peels	Chromium	Wash; Dry for 3 weeks; Grind BAT—2 g BS in 250 mL MS; [Cr(VI)] = 0.001 M; CT—5–360 min; Stir—180 rpm	Removal of up to 98% from synthetic chromium solution	[110]

Table 3. Citrus waste-derived bio-sorbents via heat/chemical and enzyme pre-treatment for removal of heavy metals from wastewater.

Bio-Sorbent Modification by Heat/Enzyme/Chemical Treatment (BPT- Bio-Sorbent Pre-Treatment)				
Adsorbent	**Heavy Metals**	**Processing Method**	**Maximum Adsorption Capacity (per Gram Adsorbent)**	**Ref.**
Citrus limetta peels	Lead	Wash; Dry—343 K, 24 h; Grind, P.S. $\leq$ 1 mm BPT—10% isopropanol, 303 K, 24 h; 1 M citric acid, 303 K, 1 h, 353 K, 1 h; 0.1 M NaOH, 303 K, 6 h, 353 K, 0.5 h; oxidation with 50% H_2O_2, 353 K, 2 h at pH 11 BAT—0.1 g PTBS in 50 mL MS; $[M^{2+}]$ = 100 mg/L; pH = 4.0, T = 303 K; RSs—100 rpm; CT—6 h	630 mg/g adsorbent. Cold alkali treatment increases adsorption by 87% (80% in first 15 min)	[96]
Citrus tamurana Citrus latifolia peels	Nickel, Cadmium, Lead from *Oriza sativa* (rice)	Wash; Sundry; Grind, P.S. = 250 μm; BPT—(a) Soaking in 1% *w/v* citric acid for 10 min, drained, dried at 423 K, 24 h CTBS (citric acid-treated BS) (b) CTBS heated to 673 K, powdered BSAC (BS active carbon) (c) BSAC treated with 1% *w/v* phosphoric acid, dried and sieved ACPA (active carbon treated with phosphoric acid) BAT—0.1 g CTBS, BSAC and ACPA added to 5 g of raw and rinsed rice, soaked in 250 mL DI with 2% NaCl at pH 6.3, 298 K, 1 h	Rice soaked with ACPA showed maximum reduction in heavy metal concentration Cd is reduced by 96.4%, Ni by 67.9%, Pb by 90.11%	[111]
Orange, Grapefruit peels	Cadmium	Wash; Dry; Grind, P.S. = 1–1.1 mm; BPT—Protonation —20 g BS in 1 L of 0.1 M HNO_3, 240 min stirring, rinsed with DI, dried at 313 K for 740 min, protonated BS BAT—0.05 g protonated BS in 50 mL MS; $[Cd^{2+}]$ = 10–1000 mg/L (0.089–0.89 mM); CT—180 min; pH 5.0	Adsorption of >90% Cd in 50 min. Desorption of the bio-sorbent material using 0.1 M HNO_3 + 0.1 M $Ca(NO_3)_2$ shows 90% recovery in 60 min	[112]
Orange peels	Cadmium	Wash; Dry; Grind, P.S. = 1–1.1 mm BPT—Protonation—10 g BS in 500 mL 0.1 N HNO_3, stirred for 4 h at 120 rpm, 298 K, rinsed with DI till pH 4.0, dried at 318 K for 12 h; PS = 1–2 mm DAT—Acrylic bed column; length 30 cm, diameter 1.3 cm; packed with 5.0 g Protonated BS; Bed height = 24–75 cm; $[Cd^{2+}]$ = 5–15 mg/L; FR = 2–15.5 mL/min; pH 5.5; T = 298 K	0.40 mmol/g adsorbent	[97]

Table 3. *Cont.*

Bio-Sorbent Modification by Heat/Enzyme/Chemical Treatment (BPT- Bio-Sorbent Pre-Treatment)				
Adsorbent	Heavy Metals	Processing Method	Maximum Adsorption Capacity (per Gram Adsorbent)	Ref.
Orange juice residue waste	Arsenate, Arsenite	Wash; Vacuum dry; Grind, P.S. = 208 μm BPT—Step I—Decolorization: 10 g BS in 500 mL in 80% EtOH, stirred for 24 h, 298 K, filtered, washed with EtOH until colorless Vacuum-dried for 24 h, DBSG (de-colorized bio-sorbent gel) Step II—Cross-linking: 5.22 g DBSG, stirred with 200 mL DMSO for 24 h, 298 K; added 40 mL epichlorhydrin, stirred for 2 h; added 50 mL of 5 M NaOH, stood for 5 h at 323 K; cooled, filtered and washed with 70% EtOH, 0.5 M HCl, again with EtOH to pH 7.0; vacuum-dried, 24 h, cross-linked BS Step III—Phosphorylation: cross-linked BS soaked in 200 mL DMF overnight, filtered and immersed again in 200 mL DMF+ 5.04 g urea; added 3.1 g phosphoric acid drop-wise with constant stirring; stirred for 1 h. Temperature raised to 423 K and stirred for 2 h; cooled to RT, filtered, washed with 70% EtOH and DI until pH 7.0; washed with 0.1 M HCl and DI until pH 7.0; vacuum-dried for 2 days. Phosphorylated BS Step IV—Fe(III) loading: treated with Fe(III) solution of concentration 55.85 mg/L (= 1 mM), Fe(III)-loaded BS BAT—25 mg Fe(III)-loaded BS in 15 mL MS ([Arsenite/Arsenate] = 15 mg/L, 24 h, T = 303 K) DAT—column packed with 0.1 g Fe(III)-loaded BS and conditioned with pH 4 water overnight. [Arsenate/Arsenite] = 15 mg/L; FR—0.098 mL/min	Bio-sorbent pre-treatment and Fe(III) loading enables direct removal of arsenite and arsenate together, without oxidizing arsenite into arsenate DAT—0.91 mmol/g adsorbent; 99% removal of arsenic compared to 80% removal by cellulose control DAT—Maximum arsenic adsorbed on the packed bed = 1.1 mg. Elution with 0.1 M HCl recovers 0.62 mg arsenic (60% recovery)	[113]
Orange peels	Cadmium, Copper, Lead	Wash; Sun dry, 6 days; Grind, P.S. = 0.2 mm BPT—Protonation—10 g dried peel soaked in 1 L of 0.1 M HNO₃, 6 h; filtered, rinsed with DI; sun-dried for 6 days. Protonated BS BAT—0.1–1 g protonated BS in 25 mL MS; [M²⁺] = 20 mg/L; CT—5–120 min; T 98 K; pH 5.0. Shaking at 200 rpm	Maximum adsorption of Pb = 73.53 mg/g; Cu = 15.27 mg/g; Cd = 13.7 mg/g. Pb(99.5%) > Cu(89.57%) > Cd(81.03%) at [M²⁺] = 20 mg/L and BS loading of 4 g/L Pb(96.3%) > Cu(93.3%) > Cd(85%) at [M²⁺] = 100–600 mg/L	[114]

Table 3. *Cont.*

Bio-Sorbent Modification by Heat/Enzyme/Chemical Treatment (BPT- Bio-Sorbent Pre-Treatment)				
Adsorbent	**Heavy Metals**	**Processing Method**	**Maximum Adsorption Capacity (per Gram Adsorbent)**	**Ref.**
Orange peels	Lead, Cadmium, Zinc	Wash; Dry; Grind, P.S. = 1–2 mm; BPT—Protonation—10 g BS in 500 mL 0.1 N HNO_3, stirred for 4 h at 120 rpm, 298 K, rinsed with DI until pH = 4.0, dried at 318 K for 12 h; PS = 1–2 mm, protonated BS DAT—Acrylic column; length 30 cm, diameter—1.3 cm, 5 g protonated BS wet-packed. Feed concentration $[Pb^{2+}]$ = 10.36 mg/L; $[Cd^{2+}]$ = 5.62 mg/L; $[Zn^{2+}]$ = 3.27 mg/L. Total feed to the column = 20 L; FR = 9 mL/min, pH 5.0, T = 298 K	Pb (85 mg/g) > Cd (44 mg/g) > Zn (20 mg/g)	[98]
Citrus paradisi (Grapefruit) peels	Zinc, Nickel	Wash; Dry, 323 K until constant weight; Grind, P.S. 0.5–1.0 mm BPT—(a) Blocking of –COOH group—9.0 g BS suspended in 633 mL CH_3OH, and 5.4 mL HCl; stirred at 100 rpm, 6 h. Centrifuged, washed, freeze-dried. (b) Blocking of –OH group—5.0 g BS suspended in 100 mL HCHO, stirred at 100 rpm, 6 h. Centrifuged, washed and freeze-dried BAT—100 mg BS in 100 mL MS; $[M^{2+}]$ = 300 mg/L, pH 5.0; CT = 120 min; SRs—100 rpm; 298 K	Native peel BS Ni^{2+}—1.331 meq/g (84.73%) Zn^{2+}—1.512 meq/g (92.46%) -COOH blocking reduces Ni^{2+} sorption by 78.57%, Zn^{2+} sorption by 73.31% -OH blocking reduces Ni^{2+} sorption by 22.63% and Zn^{2+} sorption by 28.54%	[69]
Citrus peel pectin	Lead	Citrus peel pectins (a) Low methoxylated (LM) pectin (methoxyl content 9%) and (b) high methoxylated (HM) pectin (methoxyl content 64%) BAT/KS—0.02 g BS in 200 mL MS; $[M^{2+}]$ = 0.1–1.0 mM; pH ≤ 5.0; T 294–298 K; CT—2–1440 min. Background electrolyte concentration—0.01 M $NaNO_3$	LM Pectin —0.86 mmol/g HM Pectin —0.87 mmol/g	[115]
Orange waste	Phosphate	Metal-loaded orange waste bio-sorbent: La(III)-loaded, Ce(III)-loaded and Fe(III)-loaded BS BAT—25 mg of La(III)- and Ce(III)-loaded BS and 60 mg of Fe(III)-loaded BS in 15 mL phosphate solution: (Phosphate) = 20–40 mg/L; pH = 7.5 for La(III)/Ce(III)-loaded BS and 3.0 for Fe(III)-loaded BS experiments. SRs = 140 rpm, 24 h, 303 K DAT—Glass column length—20 cm, diameter—0.8 cm; loaded with 150 mg of wet metal-loaded BS, (Phosphate) = 20–40 mg/L; FR = 7 mL/h	Phosphate adsorption by M—loaded BS (% removal) in BAT La(III)-loaded BS —13.84 mgP/g (98.5%) Ce(III)-loaded BS —14.0 mgP/g (98.8%) Fe (III)-loaded BS (99% removal) 13.63 mgP/g adsorbent in DAT	[116]

Table 3. *Cont.*

Bio-Sorbent Modification by Heat/Enzyme/Chemical Treatment (BPT- Bio-Sorbent Pre-Treatment)				
Adsorbent	**Heavy Metals**	**Processing Method**	**Maximum Adsorption Capacity (per Gram Adsorbent)**	**Ref.**
Orange waste	Nickel, Cobalt, Cadmium, Zinc	Wash; Dry, 323 K, 72 h. Ball mill—P.S. 0.1–0.2 mm; pore size—30.5 Å, BET surface area—128.7 m^2/g BPT—Different treatments, viz., isopropyl alcohol, alkali saponification, acid oxidation to yield OP, PA, SNa, Sam/SCa, SOA, SCA, SPA; BAT—0.025 g BS in 15 mL MS; [M^{2+}] = 0.001–0.01 M, CT—3 h	SPA—Ni^{2+}—1.28 mol/Kg (95% increase) SPA—Co^{2+}—1.23 mol/Kg (178% increase) SCA—Cd^{2+}—1.13 mol/Kg (60% increase) SOA—Zn^{2+}—1.23 mol/Kg (130% increase) in comparison to raw orange peel (OP) Zn^{2+} → SCA>SNa> SOA>SPA> Sam>SCa>OP Cd^{2+} → SOA>SCA>SPA>SNa> Sam>SCa>OP Co^{2+}/Ni^{2+} → SPA> SCA>SOA>SNa> Sam>SCa>OP	[85]
Orange peel	Lead, Zinc, Copper	Wash; Dry, 333.15 K; Grind: P.S. $\leq$ 0.45 mm –(OP); BET—0.828 m^2/g BPT—100 g dried OP + 500 mL EtOH + 0.8 M NaOH + 0.8 M CaCl$_2$; soak for 20 h, filter, wash until neutral pH—SCOP; BET—1.496 m^2/g BAT—0.1 g BS (OP and SCOP) in 25 mL MS; [Pb^{2+}] = 200 mg/L; [Zn^{2+}] = 50 mg/L; [Cu^{2+}] = 50 mg/L; SRs—120 rpm; CT—0–12 h; T 298 K	Adsorption capacity SCOP/OP (mg/g) Cu^{2+} → 70.73/44.28 Pb^{2+} → 209.8/113.5 Zn^{2+} → 56.18/21.25 Maximum adsorption was found at pH 5.5 Pb^{2+}(99.4%) > Cu^{2+}(93.7%) > Zn^{2+}(86.6%)	[117]
Orange Peels	Chromium	Wash; Dry, 353 K; Grind, PS $\leq$ 200 μm BPT—100 g OP + 1 L of 0.1 M NaOH. Soak for 48 h; shake at 120 rpm, filter, wash, dry at 353 K—MOP (BET—0.8311 m^2/g) BAT—0.2–5.0 g MOP in 50 mL MS; [Cr(VI)] = 100 mg/L; SRs—120 rpm; T 298 K; CT = 30 min–4 h; pH 1–8.0	OP → 97.07 mg/g (39.9%) AOP → 139.0 mg/g (41.4%) Maximum adsorption was found at pH 2.0 and BS dose of 4 g/L in 180 min	[118]
Citrus lemon	Cobalt	Wash; Dry, 353 K, 24 h; Grind BPT—Thermal activation in air at 773 K, 1 h. Wash; dry, 373 K, 24 h. PS: BS 150–200 BAT—10 g/l BS; [Co^{2+}] = 0–1000 mg/L; CT—10 h; SRs—200 rpm; pH 6.0	22 mg/g adsorbent	[119]
Orange waste Peel (OP), Bagasse (OB), Peel-bagasse (OPB)	Lead	Wash; Dry, 333.15 K; Grind, Sieve—BS—100-mesh BPT—1 g BS + 20 m 0.1 M NaOH, agitation—2 h; Wash, dry—328 K, 24 h. 1 g modified BS + 8.3 mL 1.2 M citric acid; agitation—30 min; filter; dry—328 K, 24 h; heat—393 K, 90 min; Wash; dry-328 K, 24 h BAT—0.5 g BS (OP, OB, OPB)/modified BS (OMP, OMB, OMPB) in 50 mL MS; [Pb^{2+}] = 700 mg/L; pH 2.0–6.0; T = 303 K; CT—10–1440 min	Highest adsorption capacity shown by O-MP—84.53 mg/g OP—55.52 mg/g OB—46.90 mg/g OMB—80.19 mg/g OPB—32.55 mg/g OMPB—73.37 mg/g	[120]

Table 3. *Cont.*

Bio-Sorbent Modification by Heat/Enzyme/Chemical Treatment (BPT- Bio-Sorbent Pre-Treatment)				
Adsorbent	**Heavy Metals**	**Processing Method**	**Maximum Adsorption Capacity (per Gram Adsorbent)**	**Ref.**
Pomelo Peels	Copper	Wash; Dry, 343 K, 2–3 h. Grind, Sieve—(Pomelo Peel, PP) BPT—PP+ hot acidified water, pH 1.5, T = 388–393 K, 60 min; Filter. The solid peel residue after filtration is de-pectinated pomelo peel or DPP. Washed and dried at 343 K, 2–3 h; Grind: PS—0.42 mm. Filtrate contains pectin, precipitated by 95% EtOH. BAT—0.5 g BS (PP, DPP) + 100 mL MS; $[Cu^{2+}]$ = 25 mg/L; pH = 2.0–6.0; SRs—150 rpm; T 298 K; CT—180 min	Pomelo peel (PP)—19.7 mg/g De-pectinated pomelo peel (DPP)—21.1 mg/g	[121]
Lemon Peel	Cadmium	Wash; Dry—323 K; Grind: PS—0.5–1.0 mm—native peel (NP) BPT—(a) Protonation—10 g NP + 1.0 L of 0.1 M HCl, stir—6 h, 120 rpm; T 298 K; Filter, wash; Dry—323 K—protonated peels (PrP) (b) 10 g NP + 400 mL diluted HCl; pH 1.5 at 358 K; Stir—150 rpm—2 h; T 358 K. Filter: Filtrate is coagulated by 95% EtOH. Wash with 60%, 80% and 95% EtOH with a retention time of 30 min in each washing; vacuum dry—323 K—peel pectic acid (PP) (c) The solid residue left after pectic acid extraction is washed to remove all soluble sugars; vacuum dry—323 K—De-pectinated peels (DPP) BAT—50 mg BS (NP, PrP, PP, DPP) in 50 mL MS; $[Co^{2+}]$ = 100 mg/L (1.78 mequiv./L); pH 5.0; SRs—120 rpm; T 298 K; CT—180 min	Native peel (NP)→ 1.92 mequiv./g Protonated peels (PrP) → 2.44 mequiv./g De-pectinated peels (DPP)→ 1.75 mequiv./g Peel pectic acid (PP)→ 2.86 mequiv./g PP > PrP > NP > DPP	[122]
Orange peels, Lemon Peels, Lemon-based pectin peels (PP)	Cadmium	Native orange and lemon peels → Wash, Dry—311–313 K, 12 h; Grind BPT—Protonation—Lemon-based pectin peels are treated with 0.1 N HNO_3, 6 h; Dry for 12 h, 311–313 K; Wash, Dry; Grind— PS: 0.7–0.9 mm—Protonated pectin peels (PPP) BAT—0.1 g BS (NOP, NLP, PPP) in 50 mL MS; $[Co^{2+}]$ = 10–700 mg/L; Shake—6 h; pH 3.0/5.0	0.7–1.2 mequiv./g (39–67 mg/g)	[123]
Citrus Pectin forms	Lead	Different forms of citrus pectins: GA oligomers—Large DP, medium DM and small DP size class GA oligomers (galacturonic acid) PME demethylated pectin (DM—50–80%) Pectin from peel residue Non-calcium-sensitive pectin Lyophilized BS from single-state fermentation of citrus peel (hydrolysis) BAT—50 mg BS in 50 mL MS; $[Pb^{2+}]$ = 0.5–1 g/l; pH 4.5; SRs—120 rpm; T 298 K; CT—6 h	Medium DP size class GA—380 mg/g, small DP size class GA—360 mg/g Large DP size class GA—300 mg/g PME demethylated pectin—220–270 mg/g Pectin from peel residue- 140 mg/g Non-calcium-sensitive pectin (NCSP)—200 mg/g Lyophilized BS from fermented citrus peel—100 mg/g	[89]

Table 3. *Cont.*

Bio-Sorbent Modification by Heat/Enzyme/Chemical Treatment (BPT- Bio-Sorbent Pre-Treatment)				
Adsorbent	**Heavy Metals**	**Processing Method**	**Maximum Adsorption Capacity (per Gram Adsorbent)**	**Ref.**
Orange Peels	Copper Cadmium	Wash; Dry—343 K, 24 h; Grind: PS—0.45 mm—native peel (NP) BPT—(a) 60 g NP + 300 mL 1% NaOH + 300 mL EtOH; RT; 24 h. Wash; Dry, 343 K, 24 h, DPOP (De-pigmented orange peels) (b) 30 g DPOP + 1 L 1% merceptoacetic acid, 12 h; Wash; Dry—343 K, 12 h. Grind—PS $\geq$ 0.45 mm BAT—50 mg BS in 10 mL MS; $[M^{2+}]$ = 0.05–1 g/l; SRs—120 rpm; CT—1.5 h; T 298 K, pH—5.0–7.0	Cu^{2+}—70.67 mg/g Cd^{2+}—136.05 mg/g	[86]
Lemon Peel	Cobalt	Wash; Dry—333.15 K, 24 h. Grind: PS= 1 mm—native peel (NP) BPT—10 NP + 100 mL 2% IPA, 0.1 N NaOH, 0.1 N HCl, 0.1 N H_2SO_4, 0.1 N HNO_3; 4 h, 303 K, Wash; Dry—333.15 K, 24 h BAT—0.1 g BS in 50 mL MS; $[Co^{2+}]$ = 100 mg/L, T 303 K, SRs—150 rpm; CT—6 h	Native Peels—20.83 mg/g Modified Peels—35.7 mg/g	[124]
D-limonene	Mercury	Direct reaction between sulfur and D-limonene at (a) 443 K, 1 h; (b) 453 K, 50 mm Hg, 4 h; (c) 373 K < 1 mm Hg, 5 h	55% removal	[125]

Table 4. Citrus waste reuse as bio-sorbents for the removal of poisonous dyes from wastewater.

Adsorbent	Dye	Processing Method	Maximum Adsorption Capacity (Per Gram Adsorbent)	Ref.
Citrus reticulata	Acid Yellow-73	Wash; Sun dry—7 days; Grind; Sieve through 50 ASTM mesh BPT-Soak—10% formaldehyde; air-dried—3 days; Oven-dried—353 K, 2 h BAT—1.0 g BS in 50 mL dye solution; (Dye) = 20 ppm; pH = 3.0; T = 323 K; SRs—100 rpm; CT—65 min;	96.46 mg/g^{-1}L^{-1}	[75]
Citrus sinensis	Congo Red, Rhodamine B, Procion orange	Wash; Sun dry—7 days; Grind—PS= 75–500 μm BAT—250 mg BS in 50 mL Congo Red dye solution; (Congo Red) = 60 mg/L; CT—20–90 min, SRs—140 rpm; T 302 K; pH 5.0 500 mg BS in 50 mL Rhodhamine B and Procion orange dye solutions; (Dye) = 10 mg/L; CT—20–90 min, SRs—140 rpm; T 302 K; pH 3.0	Congo Red—22.4 mg/g; pH = 5.0 (76.6%) Procion orange—1.3 mg/g; pH = 3.0 (49%) Rhodamine B—3.22 mg/g; pH = 3.0 (38.43%)	[126]
Grapefruit peels	Methylene Blue	Wash; Sun dry—2 days; Grind—PS > 90 μm BPT- Carbonization—Treat with (a) BS_ 88% orthophosphoric acid (1: 3 ratio), (b) $ZnCl_2$, or (c) 98% H_2SO_4 Heat at 723–823 K for 0.75–1.5 h, wash with NH_4OH and H_2O to neutral pH, Dry—12 h; Charred citrus peel (CCF); PS= 135 μm BAT—0.30–1.0 g CCF in 200 mL MB dye solution; (MB) = 20–100 mg/L; T 303 K; CT—8 h; pH 3.0–10.0	99.08% removal	[72]

Table 4. *Cont.*

Adsorbent	Dye	Processing Method	Maximum Adsorption Capacity (Per Gram Adsorbent)	Ref.
Orange Peel	Direct Yellow-12	Wash; Dry—423 K, 5 h; Grind BPT—Carbonization—3 kg dried orange peel + 2.5 L 98% H_2SO_4, Stand—2 h; Boil—3 h; Add to ice-cold water, filter; Dry—453 K, 2 h; immerse in 5.0 L of 5% $NaHCO_3$, wash to neutral pH; Dry—423 K, 3 h; Grind $\leq$ 0.200 mm BAT—0.5 g in 100 mL dye solution; (Dye) = 75 mg/L; T 300 K; CT—2 h; pH 1.5–11.2; SRs—200 rpm	96% removal	[127]
Grapefruit Peels	Crystal Violet	Wash; Dry—423 K, 5 h; Grind BAT—BS = 0.1–3 g/L of dye solution; (Dye) = 5–600 mg/L; pH 6.0; SRs—100 rpm; T 303 K; CT—60 min	96% removal in 60 min. Maximum adsorption capacity = 254.16 mg/g	[128]
Pomelo Peel	Methylene Blue (Cationic Dye); Acid Blue (Anionic Dye)	Wash; Air dry; Grind-PS: 1.0–2.0 mm BPT—Microwave modification: BS + 1:1.25 by wt. NaOH. Microwave heating at 2.45 GHz, 800 W, 5 min; Wash with 0.1 M DI until neutral pH BAT—0.20 g Modified BS in 200 mL dye solution; (Dye) = 50–500 mg/L; SRs—120 rpm; T 303 K; CT—until equilibrium	Methylene Blue— 501.1 mg/g Acid Blue— 444.45 mg/g	[129]
Pomelo Peel	Congo Red	Wash; Dry—313 K, 48 h; Ball Mill-PS= 0.840 mm BAT—1.0–3.0 g BS in 1 L dye solution; (Dye) = 20–120 mg/L; T 276.15–333.15 K, pH 6.0–8.7; CT- 24 h	0.75–1.08 mg/g	[130]
Citrus medica, Citrus aurentifolia, Citrus documana	Reactive Red 2 (Red M5B)	Wash; Dry—373–393 K, 24 h; Crush BPT—Carbonization—Heat at 773 K, N_2; Liquid phase oxidation with 1 M HNO_3; Wash, Dry—423 K, 12 h BAT—3 g BS in 100 mL dye solution; (Dye) = 20 mg/L; T 298 K; CT = 5–90 min; SRs—120 rpm; pH 3.0–10.0	*C. medica* $\rightarrow$ 87% (0.5800 mg/g) *C. aurentifolia* $\rightarrow$ 85% (0.5667 mg/g) *C. documana* $\rightarrow$ 91% (0.6067 mg/g)	[70]
Citrus limonum	Methyl Orange, Congo Red	Wash; Dry—373 K, 24 h; Grind BPT—Heat at 773 K in air, 1 h; Wash; Dry—373 K, 24 h; PS-BSS 100–250 BAT—0.1 g BS in 10 mL dye solution; (Dye) = 0.3–0.45 mM; T 298 K; CT-until equilibrium; pH 5.5–6.5	Methyl Orange $\rightarrow$ 50.3 mg/g Congo Red $\rightarrow$ 34.5 mg/g	[131]
Citrus sinensis bagasse	Methylene Blue	Wash; Dry—333 K, 72 h; Grind-PS: 0.25–0.75 mm BAT—0.1 g in 100 mL dye solution; (Dye) = 50 mg/L; CT—24 h; T 303 K; pH 7.0	96.4 mg/g	[132]
Orange Peel, Lemon Peel	Methylene Blue	Wash; Dry—353 K, 24 h; Grind-PS: <3.0 mm BAT—0.25 g in 25 cm^3 dye solution; (Dye) = 50–1000 mg/dm^3; T = 298 K; pH 2.0–3.0	Orange Peel $\rightarrow$ 4.76–95.03 mg/g Lemon Peel $\rightarrow$ 4.41–92.1 mg/g	[133]
Grapefruit Peel	Leather Dye mixture: Sella Solid Blue, Special Violet, Derma Burdeaux, Sella Solid Orange	Wash; Dry—333.15 K, 24 h; Grind-PS: <0.5 mm BPT—1.5 g BS + 150 mL of 1 M H_2O_2; Stirring—110 rpm, 24 h; Dry; Grind BAT—0.3–1.5 g BS in 50 mL dye solution; (Dye) = 100–400 mg/L; pH 5.5; T 298 K; SRs—120 rpm; CT = 24 h	Untreated Grapefruit peel BS $\rightarrow$45% Modified Grapefruit peel BS $\rightarrow$ 80% Maximum capacity $\rightarrow$ 1.1003 meq/g Maximum uptake $\rightarrow$ 37.427 mg/g	[134]

Table 4. *Cont.*

Adsorbent	Dye	Processing Method	Maximum Adsorption Capacity (Per Gram Adsorbent)	Ref.
Orange peel	Congo Red, Methyl Orange	Wash; Sun dry—72 h; Grind BPT—(a) BS + 1% NaOH, EtOH; Filter; Wash, Air dry —OP (removal of lignin and pigments) (b) 10 g OP + 100 mL DI; Stir and heat at 353 K; 120 min; Cool to RT, add N-vinyl-2-pyrrolidone, stir for 5 min—NVP/OP copolymer (c) Transfer to glass tubes and irradiate with Gamma source—radiation dose (10–50 kGy); dose rate—1.46 kGy/h. Cross-linked NVP/OP hydrogel; Wash; Dry in air BAT—1:1 BS in 20 mL dye solution; (Dye) = 10–50 mg/L; T = 293–333 K; pH = 7.0 for Congo Red and 6.0 for Methyl Orange; CT (Congo Red) = 6000 min, (Methyl Orange) = 4000 min	Congo Red → 4.8–26 mg/g Methyl Orange → 4.6–10 mg/g	[73]
Citrus grandis	Methylene Blue	Wash; Dry—333.15 K, 48 h; Grind: PS= 0.5–1.0 mm BAT—0.20 g BS in 200 mL dye solution; (Dye) = 50–500 mg/L; pH 7.0; T 303 K; CT = 5.15 h; SRs—100 rpm	344.83 mg/g at 303 K	[135]
Mosambi peels	Erichrome Black T	Wash; Sun dry; Grind; Dry—333 K, 24 h BPT—BS + Concentrated H_2SO_4 (1:1)- 24 h; Dry—378 K, 12 h; Wash with $NaHCO_3$; Dry—378 K, Mosambi peel activated carbon (MPAC)	93.8%	[136]
Citrus sinensis L.	Remazol Brilliant Blue	Wash, Dry—333 K, 24 h; Grind: Wash, Dry—333 K; PS = 44–1180 μm BAT—300 mg BS in 30 mL dye solution; (Dye) = 30, 100, 250 mg/L; SRs = 150 rpm; T 293–333 K; CT = 24 h	11.62 mg/g	[137]
Citrus sinensis	Reactive Blue 19, Reactive Blue 49	Wash; Dry; Grind—PS < 0.25 mm—BS BPT—(a) Immobilization: BS + sodium alginate (1:2). The resultant beads preserved in 0.02 M $CaCl_2$ solution. Immobilized BS (b) 1 g BS + 5% glacial acetic acid. Wash after 1 h; Dry—343 K, 24 h. Acetic acid-treated BS BAT—0.5–1.5 g in 50 mL dye solution; (Dye) = 50–300 mg/mL; CT = 60–120 min; pH 2.0; T 303 K; SRs—100 rpm	Reactive Blue 19 BS → 37.45 mg/g Immobilized BS → 400.00 mg/g Acetic acid-treated BS → 75.19 mg/g Reactive Blue 49 BS → 135.16 mg/g Immobilized BS → 80.00 mg/g Acetic acid-treated BS → 232.56 mg/g	[66]
Citrus waste	Methylene Blue	Wash; Dry; Grind—383 K, 24 h; PS < 0.5 mm BAT—0.70 g in 100 mL dye solution; (Dye) = 5–60 mg/L; T 300 K, CT = 180 min; SRs—120 rpm	3.2994 mg/g adsorbent at (Dye) = 50 mg/L. Maximum removal percentage → 49.35% at 60 mg/L	[138]
Lime Peel	Remazol Brilliant Blue R	Wash; Dry—T 378 K, 24 h; PS= 150 μm BAT—1–9 g BS in 50 mL dye solution; (Dye) = 10–50 mg/L; SRs—120 rpm; CT = 24 h; T 300 K	73–95.89% removal. Adsorption capacity of 7.29–9.58 mg/g	[139]

Table 4. *Cont.*

Adsorbent	Dye	Processing Method	Maximum Adsorption Capacity (Per Gram Adsorbent)	Ref.
Pomelo Peel	Malachite Green	Wash; Dry—T 393 K, overnight; Grind BPT—Carbonization—973 K, N_2, 1 h; Char is soaked in KOH (1:1); Dry—423 K, overnight; Pyrolyze at 1073 K; N_2 (150 cm^3/min); At T = 1073 K, CO_2 flow for 2 h; Cool to RT under N_2 flux; Wash with 0.1 M HCl; Wash until neutral pH; BET—1357.21 m^2/g BAT—0.2 g BS in 100 mL dye solution; (Dye) = 50–400 mg/g; T 293–333 K; CT = 4 h; SRs = 120 rpm; pH 3.0–10.0	178.48 mg/g Best result at pH = 8.0; T = 333.15 K 95.06% removal	[140]
Citrus reticulata	Indigo Caramine Dye	Wash; Sun dry, 7 d;ays Oven dry—T 343 K, 4 h BPT—(a) 100 g BS + 500 mL 20% formaldehyde, 3 h, FBS (b) 100 g BS + 10% urea solution, 3 h, UBS (a, b)—Filter, air dry; Oven dry, 343 K, 4 h; Grind: Sieve through 50-mesh ASTM = 297 μm BAT—0.3–3.0 g BS in 50 mL dye solution; (Dye) = 25 mg/L; T 293–343 K; CT = 10–70 min; pH 1.0–10.0; SRs—100 rpm	Dried Peel (BS) → 5.90 mg/g Formaldehyde-treated peel (FBS) → 14.79 mg/g Urea-treated peel (UBS) → 71.07 mg/g	[74]
Orange Peels	Acid Violet 17	Wash; Sun dry—4 days; Grind; PS: 53–500 μm BAT—100–600 mg BS in 50 mL dye solution; (Dye) = 10 mg/L; pH 2.0–10.0; CT—80 min; T 303 K	19.88 mg/g; 87% removal at pH = 2.0 and 100% removal at pH = 6.27 at adsorbent dose of 600 mg in 50 mL of 10 mg/L dye solution	[141]
Citrus limetta Peels	Methylene Blue	Wash; Sun dry—4 days; Oven dry—363 K, 24 h; Grind; Sieve, 80 BSS mesh; Wash; Dry—T 378 K, 4 h; Grind; PS= 80–200 BSS BAT—0.05 g in 25 mL dye solution; (Dye) = 25–250 mg/L; CT—3 h; pH 4.0	227.3 mg/g; 97–98% removal	[48]
Citrus sinensis Peels	C.I. Direct Blue 77 dye	Wash; Dry—T 378 K; Grind; PS = 75 μm BAT—5–30 mg in 100 mL dye solution; (Dye) = 50 mg/L; pH 2.0–12.0; SRs = 125 rpm; CT = 60 min	59% removal; 9.43 mg/g	[113]
Orange peel	Methylene Blue	Wash; Dry—343 K, 5 h; Grind; PS= 75 μm, OP (dried orange peel) BPT—60 g OP + 250 mL of 0.1 M NaOH, 24 h; Filter; Wash; Dry—MOP (Modified OP) BAT—0.05 g in 25 mL dye solution; (Dye) = 35 mg/L; SRs = 160 rpm; pH 4.0; CT = 30 min; T 298–318 K	OP → 14.164 mg/g MOP → 18.282 mg/g	[142]

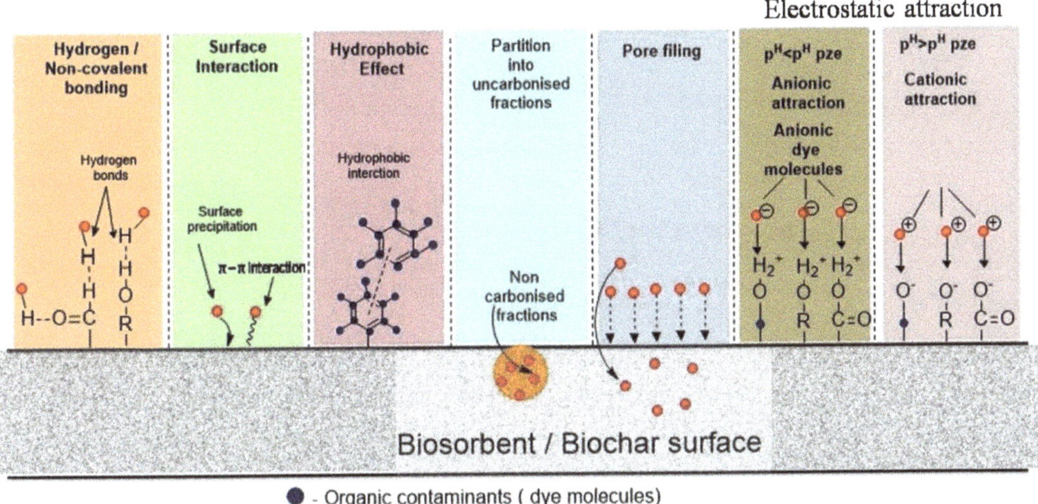

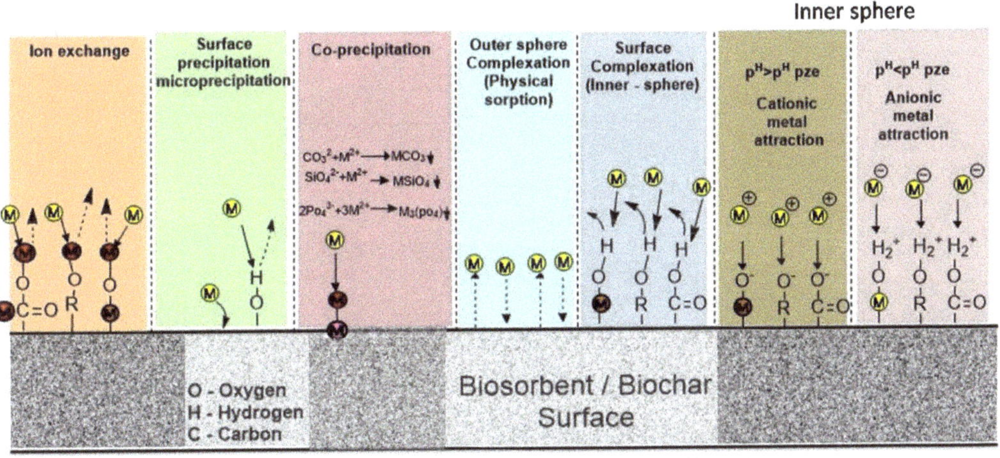

Figure 9. Mechanism of adsorption of dyes and heavy metals from industrial wastewaters by citrus biomass-derived bio-sorbent. Artwork developed from the information provided in [47].

4. Kinetics and Thermodynamics

4.1. Kinetics

The adsorption kinetics helps in defining the rate of efficiency of adsorption. Kinetic parameters are useful for designing and modeling the bio-sorption processes. Several types of kinetic models have been proposed by researchers to study the mechanism and rate-regulating steps. The Lagergren's rate equation is one of the most commonly used models to explain the adsorption of an adsorbate from the liquid phase [143]. The linear form of the pseudo-first-order equation is given as [144]:

$$\ln(q_e - q_t) = \ln q_e - k_1 t \tag{1}$$

where, 'q_t' and 'q_e' are amount of adsorbed material (mg/g) at time 't' and at equilibrium, respectively. 'k_1' (min^{-1}) is the rate constant for the pseudo-first-order reaction. The pseudo-second-order kinetics explain the involvement of both the adsorbate and adsorbent in the rate-limiting step [113,144]. The equation for the linearized pseudo-second-order reaction is given as:

$$\frac{t}{q_t} = \frac{1}{k_2\, q_e^2} + \frac{t}{q_e} \tag{2}$$

where, 'k_2' is the rate constant (g/mg^{-1} min^{-1}) of the second-order kinetics [113,145]. The initial adsorption rate 'h' (mg/g min^{-1}) is defined as:

$$h = k_2 q^2 e \tag{3}$$

the values of 'k_2' and 'h' can be obtained from the intercept of the plot based on the second-order equation [146]. The kinetics defined by Elovich's model is based on the principle that adsorption sites increase exponentially with progress in the adsorption process, suggesting multilayer adsorption [147]. The linear form of Elovich's equation is given as:

$$q_t = \frac{1}{\beta} \ln(\alpha\beta) + \frac{1}{\beta} \ln t \tag{4}$$

where, 'α' is the initial rate of adsorption (mg/g^{-1} min^{-1}) and 'β' is related to surface coverage (g/mg^{-1}). Weber and Morris proposed an intra-particle diffusion model [117,148]. The equation for the linear form of this model is given as:

$$q_t = kt^{1/2} + C \tag{5}$$

where, 'C' is the intercept and 'k' is the intra-particle diffusion constant. The value of 'k' can be calculated from the slope of the linear plot of 'q_t' vs. '$t^{1/2}$'. If intra-particle diffusion is involved in adsorption, then there would be a linear plot for of 'q_t' against '$t^{1/2}$'. In cases where the line passes through the origin, it shows that intra-particle diffusion is the rate-controlling step [132].

4.2. Thermodynamic Observations

Appropriate study and explanation of adsorption isotherms is very significant and crucial for the overall development of the adsorption mechanism and effective design of the adsorption system. It helps to explain the mechanism of interaction between adsorbate molecules with the adsorbent surface. There are many models and operational designs available to understand the batch adsorption system. The most commonly employed methods are Langmuir and Freundlich models. The Langmuir adsorption isotherm model explains monolayer adsorption equilibrium between the adsorbate and the adsorbent [149]. This model is suitable for explaining the chemisorption when there is covalent or ionic bond formation between the adsorbate and the adsorbent. Many systems followed the equation to explain the binary adsorption system. The Langmuir model in its linear form may be expressed as:

$$\frac{1}{q_e} = \frac{1}{K_L C_e q_m} + \frac{1}{q_m} \tag{6}$$

In this equation, 'q_e' is adsorption capacity (mg/g^{-1}) at equilibrium, 'C_e' is the equilibrium concentration (mg L^{-1}) of the adsorbate, 'q_m' is maximum adsorption capacity (mg/g^{-1}) and 'K_L' is the Langmuir constant (L mg^{-1}). The Freundlich isotherm explains the multi-layered adsorption phenomenon. It is applicable for reversible adsorption of the adsorbate on the surface of the adsorbent [120,150]. It states that the surface of the adsorbent should be heterogeneous in nature for multilayer adsorption [115,145]. This model states that the surface of the adsorbent has a diverse binding energy spectrum. The linear form of the Freundlich isotherm equation can be expressed as:

$$\log_{q_e} = \log K_F + \frac{1}{n}\log C_e \tag{7}$$

where, 'K_F' is the Freundlich isotherm constant ($mg^{1-1/n}\ L^{1/n}\ g^{-1}$), and shows the adsorption efficiency of per unit mass of adsorbent. The $1/n$ value expresses the heterogeneity factor.

Thermodynamic parameters, e.g., entropy change, enthalpy change and standard free energy, are signification parameters to assess and evaluate the viability of the adsorption process along with the nature of adsorption. The negative value of change in enthalpy ($\Delta H°$) shows the exothermic nature of the adsorption process, while the positive value of change in entropy ($\Delta S°$) stipulates the increased randomness of the process at the interface. It explains that the process is entropy-driven. The Gibbs free energy change of the adsorption process is subsequent to 'K_c' and given by the following equation:

$$\Delta G° = -RT\ln K_c \tag{8}$$

Here, 'K_c' can be expressed as shown in the following equation:

$$\ln K_c = \frac{\Delta S°}{R} - \frac{\Delta H°}{RT} \tag{9}$$

Standard free energy ($\Delta G°$), enthalpy change ($\Delta H°$) and entropy change ($\Delta S°$) can be determined by using the following equation:

$$\Delta G° = \Delta H° - T\Delta S° \tag{10}$$

where, 'R' is universal gas constant ($8.314\ J.mol^{-1}\ K^{-1}$), '$T$' is temperature in Kelvin and 'K_c' is equilibrium constant. Change in enthalpy ($\Delta H°$) and entropy ($\Delta S°$) can be calculated from the slope of the plot of ($\Delta G°$) vs. T. Reported values of $\Delta H°$ for physical adsorption range from -4 to $-40\ kJ.mol^{-1}$. Bhatnagar et al. has calculated thermodynamic parameters to check the adsorption nature of cobalt using lemon peel as a bio-sorbent. The value of ΔH ($-21.2\ kJ.mol^{-1}$) found in the range shows the physical adsorption [119]. The value of ΔG calculated indicates the spontaneity in the process. A thermodynamic study of the bio-sorption of methylene blue from *C. sinensis* bagasse was observed by Bhatti et al., and the calculated value of $\Delta H°$ ($51.9\ kJ/mol$) shows a similar physical adsorption in the process as a purely physical or chemical one [132]. The role of physisorption can be explained on the basis of the heat involved, which is $>40\ kJ.mol^{-1}$, whereas, for chemisorption, it is reported in a range of $80–200\ kJ.mol^{-1}$. Similar results have been reported for the adsorption of Reamzol Brilliant Blue using an orange peel adsorbent [137]. The process was efficient as the negative value of free energy denotes the feasibility of the process. Additionally, the positive values of ΔH and ΔS shown are in favor of the adsorption process. These results also show the affinity of the adsorbent towards the dye. However, the adsorption of dyes has been reported as an exothermic phenomenon in many studies. Bio-sorption of La and Ce using peels of *Citrus reticulata* was also found to be a thermodynamically feasible and spontaneous process. It is shown to be a process of endothermic nature in the temperature range of 293–323 K, and the overall entropy increases due to the exchange of the metal ions with more mobile ions [107]. Malachite green dye adsorbed by *Citrus grandis* peels revealed the change in $\Delta G°$ from -21.55 to $-24.22\ kJ.mol^{-1}$, in the temperature range of 303 to 333 K, indicating enhanced spontaneity at high temperatures [140]. Similar results were also observed during the removal of fluoride using a *Citrus limetta* peels adsorbent activated with $FeCl_3$ [56]. A thermodynamic study for the removal of methylene blue dye using *Citrus limetta* peel waste exhibited Gibbs free energy ($\Delta G°$) values in favor of the process [48] (Table 5). Kaffir lime peels were used to reduce graphene oxide to prepare reduced graphene oxide (RGO), and applied for the absorption of methylene blue [151]. Several kinetic models fitted for the citrus peel as adsorbent have been shown in Table 5.

Table 5. Thermodynamics and kinetics studies of various dyes and ions on citrus peel as adsorbents.

Citrus Peel as Adsorbent	Adsorbing Substance	T (Kelvin)	ΔG (kJ mol^{-1})	ΔH (kJ mol^{-1})	ΔS (J mol^{-1}K^{-1})	Isotherm Fitted	Kinetics Model	Ref.
Lemon peel	Cobalt	298, 318	−37.47, −38.56	−21.2	54.61	Langmuir model	Pseudo-second-order	[119]
Citrus sinensis (Musambi peel)	Methylene Blue dye (MB)	303, 308, 313, 318, 323	0.76, 2.40, 2.90, 3.90, 4.40	−51.9	−0.18	Langmuir model	Pseudo-second-order	[132]
Citrus sinensis (Orange peel)	Brilliant Blue dye (BB)	303, 313, 323, 333	−14.70, −16.11, −17.59, −18.32	22.82	124.20	Langmuir and Freundlich isotherms	–	[137]
Citrus sinensis (Orange peel)	Arsenic	293, 298, 303, 308, 313, 318	−30.18, −32.42, −33.35, −34.48, −35.06, −35.66	30.0	−0.21	Langmuir iotherm	Pseudo-second-order	[102,121]
Citrus grandis (Pomelo peel)	Malachite Green dye (MB)	303, 318, 333	−21.55, −22.89, −24.22	9.16	0.15	Langmuir isotherm	Pseudo-second-order	[140]
Citrus maxima Pomelo peel	Copper	298, 308, 318	−5.38, −4.19, −3.49	−32.18	0.09	Langmuir isotherm	Pseudo-second-order	[121]
Citrus limetta (Musambi peel)	Methylene Blue dye (MB)	293, 303, 313, 323	−7.87, −9.38, −10.49, −12.41	35.13	146.70	Langmuir isotherm	Pseudo-second-order	[48]
Citrus limetta (Musambi peel)	Fluoride	298, 308, 318	−0.69, −3.45, −5.61	72.58	246.22	Langmuir isotherm	Pseudo-second-order	[56]
Citrus limetta (Musambi peel)	Chromium	303, 313, 318, 323, 328	−420.21, −725.16, −726.24, −2044.33, −1151.51	1.914×10^{-3}	57.44	D–R isotherm	Pseudo-second-order	[152]
rGO-Kaffir Lime Peel Extract	Methylene Blue dye (MB)	303, 313, 333	−5.98, −6.81, −8.47	19.15	82.93	Langmuir isotherm	Pseudo-second-order	[151]
Magnetite Orange peel	Crystal Violet	303, 313, 323, 333	−5.92, −6.95, −8.02, −9.35	28.02	111.60	Langmuir isotherm	Pseudo-second-order model	[153]
Lemon peels impregnated with phosphoric acid	Erythrosine-B (EB) Rhodamine-B (RB)	298 298	−3.19 −2.97	27.43 24.41	102.76 91.87	Langmuir model	Pseudo-second-order model	[154]
Orange (*citrus sinensis*) peels by acid activation	Methylene Blue dye (MB)	313, 323, 333	−20.5, −12.9, −15.4	67.60	281.70	Langmuir model	Pseudo-second-order model	[155]

5. Design of Experiments

Response surface methodology (RSM) is a very popular tool for the optimization of process variables. It has been adopted in various studies for the design and analysis of the experiments. Principally, it is a mathematical and statistical technique for the design of experiments using relations between a cluster of controlled experimental variables and the measured properties, created on one or more selected conditions [156]. Numerical or physical experimental data are calculated by an expression that is generally a low-order polynomial. Usually, a second-order polynomial equation is fitted to analyze the experimental data by means of RSM, which can be represented as:

$$Y = b_0 + \sum_{i=1}^{n} b_i x_i + \left(\sum_{i=1}^{n} b_{ii} x_i \right)^2 + \left(\sum_{i=1}^{n-1} \sum_{j=i+1}^{n} b_{ij} x_i x_j \right) + \varepsilon \tag{11}$$

The results are obtained as 2D contours and 3D plots. This method is very competent, and uses the experimental data and interactions between the factors [157,158]. This process is based on three key steps, which involve statistically designed experiments, determination of the coefficients through estimation of response via mathematical modeling and investigating the competency of the model [159]. The ANOVA program is used to calculate the statistical parameters along with the optimization of independent parameters and dependent output responses. Dutta et al. analyzed the result of each run and correlated the responses with three individual factors for preparation of an adsorbent using an empirical second-degree polynomial, as shown above. Optimized conditions obtained as responses for carbonization of citrus fruit peel were weight ratio of the peel to the activating agent, temperature of carbonization and time of carbonization, which have the values of 3:1, 798 K and 0.75 h, respectively (Table 5) [72]. An experimental design for the removal of MB dye by charred citrus fruit peel has also been attained. Numeric parameters selected were initial concentration of MB, amount of adsorbent and pH of the solution, and results obtained from statistical design were maintained during the experiment and found to be fitted for the removal of dye [72]. The model adequacy of the Cr(VI) adsorption by Musambi peels was also found to be statistically viable [152]. In recent years, RSM modeling has been applied and reported in several adsorbent-based materials. It is an efficient and useful procedure which can help maximize the performance along with responses based on combinations of variables. Therefore, RSM offers an extensive scope for modeling the parameters' optimization along with the percent removal of heavy metal/dye by citrus peel waste-based adsorbent materials. Some of the experimental models applied for citrus peel adsorbents are recorded in Table 6.

Table 6. Applications of RSM in the bio-sorption process.

Object of Experimental Design	Independent Variables	Response(s)	Remarks	Ref.
Parameters optimized for the preparation of adsorbent from citrus fruit peel	Weight ratio (citrus fruit peel to activating agent), temperature of carbonization	Operating parameters for carbonizing of citrus peel	Optimized conditions for carbonization of citrus fruit peel are: weight ratio of peel to activating agent (3:1) at temperature of 798 K, and time of carbonization was 0.75 h	[72]
Citrus fruit peel used in the removal of methylene blue (MB) dye	Initial concentration of MB, weight of CCFP and pH	Percentage removal of MB	99.6225% MB removal at pH 3.64, weight of CCFP. Initial concentration of MB kept constant at 0.65 g and 20 mg/L. Prepared adsorbent is superior in terms of its porosity.	[72]
Citrus limetta peel dust used for removal of Cr(VI)	Initial concentration and pH of solution	Cr(VI) adsorption by musambi peel	Initial concentration 6.75, pH 4.29, dose 0.27 g/100 mL and contact time 56.40 min	[152]
Adsorption of Brilliant Green (BG) dye by adsorbent prepared from *Citrus limetta* peel	Temperature, pH, adsorbent dosage and contact time	Percentage removal efficiency of BG	The model validations as optimum levels of the process parameters to obtain the maximum adsorption of dye of 85.17% at 313 K, pH 9, at an adsorbent dose of 3.5 g/L of aqueous dye solution and contact time of 240 min	[160]
Adsorption of Eosin Y by the activated carbon (WCAC) prepared from waste citrus peel	Concentration of Eosin Y, temperature and the adsorbent dose	Adsorption of Eosin Y	Maximum dye uptake of 59.3 mg/g at the dye concentration of 50 mg/L, temperature 333 K and the adsorbent dose of 0.1056 g	[161]
Adsorption of antibiotic Trimethoprim studied by activated carbon prepared from waste citrus peel (WCAC)	Concentration of solution, pH, temperature and adsorbent dose	Adsorption efficiency of trimethoprim by WCAC	Maximum adsorption amount of TMP by WCAC calculated was 144.9 mg/g at 293 K	[162]

6. Summary and Conclusions

Biotransformation of citrus waste into valuable compounds and adsorbent substrate materials for the purpose of adsorption of heavy metals, dyes and toxic chemicals from industrial wastewaters is among hugely adopted research projects around the world. Hazards of pollution in water are not only restricted to the aquatic ecosystem but are also found to spread to the underground water tables, crops and crop products, human/livestock/birds' health and microbial ecosystems on the land. Adsorbents from citrus wastes can be developed by a number of methods, namely physical processes, chemical methods, thermal and thermo-chemical techniques. Protonated adsorbents have demonstrated efficiencies better than native peel bio-sorbents. In addition, chemically treated bio-sorbents exhibit several advantages, such as greater chemical and mechanical stabilities in the test solution. Furthermore, they help in improving the surface properties with additional functional groups or active adsorption sites and enhance the adsorptive capacities of the resultant sorption materials. Thermochemical activation gives rise to activated carbon materials, which show enhanced porosity to facilitate physisorption along with chemical adsorption. The sorption process is governed by a number mechanisms, such as physical adsorption on the adsorbent surface by van der Waals forces of attraction, hydrogen bonding, dipole-induced dipole moments and electrostatic attraction between charged species, i.e., cationic charges on heavy metal ions and polyanionic charges on the bio-sorbent surface. Two main methods of carrying out sorption processes are popular and widely adopted in experiments: batch adsorption tests and fixed-bed adsorption columns. The latter has an advantage of installing an additional number of columns in order to increase the length of the adsorption bed for enhanced capacity of removal of pollutants from wastewater. Theoretical studies, including kinetics, thermodynamics, simulation and modeling, add a greater in-depth understanding of the adsorption mechanism. Bio-sorbents derived from citrus wastes, the largest fruit crop grown on the planet, provide an inexpensive, natural, renewable and sustainable means of obtaining resourceful as well as fruitful products.

Author Contributions: This work was completed with the contributions of 8 authors. N.M. and P.A. designed and wrote the manuscript; M.S. and A.D. contributed to analyzing recent studies on the valorization of citrus waste and synthesis of bio-adsorbents from citrus peel waste; P.A. compiled the theoretical studies reported in this subject area; B.P. and D.M. contributed to the summary and interpretation of reports and relevance to the current research progress issues; M.K.T. and S.A. performed the final proof-reading of the manuscript; N.M. carried out the final editing, revision and supervision of the project to bring it to its final format. All authors have read and agreed to the published version of the manuscript.

Funding: Authors acknowledge support from the National Research Foundation of Korea (NRF) Korean Government (Ministry of Science and ICT) (NRF-2020R1G1A1015243).

Institutional Review Board Statement: Not applicable.

Informed Consent Statement: Not applicable.

Data Availability Statement: Not applicable.

Conflicts of Interest: The authors declare no conflict of interest.

References

1. Mahato, N.; Sharma, K.; Sinha, M.; Baral, E.R.; Koteswararao, R.; Dhyani, A.; Cho, M.H.; Cho, S. Bio-sorbents, industrially important chemicals and novel materials from citrus processing waste as a sustainable and renewable bioresource: A review. *J. Adv. Res.* **2020**, *23*, 61–82. [CrossRef] [PubMed]
2. Mahato, N.; Sinha, M.; Sharma, K.; Koteswararao, R.; Cho, M.H. Cho Modern Extraction and Purification Techniques for Obtaining High Purity Food-Grade Bioactive Compounds and Value-Added Co-Products from Citrus Wastes. *Foods* **2019**, *8*, 523. [CrossRef] [PubMed]
3. Mahato, N.; Sharma, K.; Sinha, M.; Cho, M.H. Citrus waste derived nutra-/pharmaceuticals for health benefits: Current trends and future perspectives. *J. Funct. Foods* **2018**, *40*, 307–316. [CrossRef]
4. Mahato, N.; Sharma, K.; Nabybaccus, F.; Cho, M.H. Pharma-/Nutraceutical Applications. *Eras J. Med. Res.* **2016**, *3*, 20–32.

5. Mahato, N.; Sharma, K.; Rao, R.K.; Sinha, M.; Baral, E.R.; Cho, M.H. Citrus essential oils: Extraction, authentication and application in food preservation. *Crit. Rev. Food Sci. Nutr.* **2017**, *59*, 611–625. [CrossRef] [PubMed]

6. Sharma, K.; Mahato, N.; Cho, M.H.; Lee, Y.R. Converting citrus wastes into value-added products: Economic and environ-mently friendly approaches. *Nutrition* **2017**, *34*, 29–46. [CrossRef]

7. Sharma, K.; Mahato, N.; Lee, Y.R. Extraction, characterization and biological activity of citrus flavonoids. *Rev. Chem. Eng.* **2019**, *35*, 265–284. [CrossRef]

8. Mahato, N.; Sharma, K.; Sinha, M.; Dhyani, A.; Pathak, B.; Jang, H.; Park, S.; Pashikanti, S.; Cho, S. Biotransformation of Citrus Waste-I: Production of Biofuel and Valuable Compounds by Fermentation. *Processes* **2021**, *9*, 220. [CrossRef]

9. Crowe, M.; Nolan, K.; Collins, C.; Carty, G.; Donlon, B.; Brøgger, M. *European Environment Agency Topic Report: Biodegradable Municipal Waste Management in Europe*; EEA: Copenhagen, Denmark, 2002; p. 123.

10. United States Department of Agriculture. Chemistry and Technology of Citrus. In *Citrus Products Agriculture Handbook*; USDA: Washington, DC, USA, 1956; Volume 98, pp. 6–24.

11. O'Neal, B.F. *Citrus Waste Research Project FLA*; State Bd Health Bur. Sanit. Engin. Final Rpt; USDA: Washington, DC, USA, 1953; p. 82.

12. Annonymous. Grappling with that problem of citrus waste odors. *Food Eng.* **1954**, *26*, 145.

13. Sanborn, N. Spray irrigation as a means of cannery waste disposal. *Cann. Trade* **1952**, *74*, 17–21.

14. Ingols, R. The citrus canning disposal problem in Florida. *Sew. Work. J.* **1945**, *17*, 320–329.

15. McNary, R.R. Citrus Canning Industry. *Ind. Eng. Chem.* **1947**, *39*, 625–627. [CrossRef]

16. Sanborn, N. Nitrate treatment of cannery wastes. *Fruit Prod. J. Am. Vineg. Ind.* **1941**, *20*, 207–208.

17. Hooper, C.H. Citrus Fruits. *Nature* **1915**, *96*, 58–59. [CrossRef]

18. Wakefield, J. The results of research on citrus waste disposal. *Proc. Fla. State Hortic. Soc.* **1953**, 246–254.

19. Murdock, D.I.; Allen, W.E. Germicidal Effect of Orange Peel Oil and D-Limonene in Water and Orange Juice. *Food Technol.* **1960**, *14*, 441–445.

20. Subba, M.S.; Soumithri, T.C.; Rao, R.S. Antimicrobial Action of Citrus Oils. *J. Food Sci.* **1967**, *32*, 225–227. [CrossRef]

21. Ahmaruzzaman, M. Industrial wastes as low-cost potential adsorbents for the treatment of wastewater laden with heavy metals. *Adv. Colloid Interface Sci.* **2011**, *166*, 36–59. [CrossRef]

22. Alshabanat, M.; Alsenani, G.; Almufarij, R. Removal of crystal violet dye from aqueous solutions onto date palm fiber by ad-sorption technique. *J. Chem.* **2013**, *2013*, 1001. [CrossRef]

23. Brusick, D. Genotoxicity of Phenolic Antioxidants. *Toxicol. Ind. Health* **1993**, *9*, 223–230. [CrossRef]

24. Dich, J.; Zahm, S.H.; Hanberg, A.; Adami, H.O. Pesticides and cancer. *Cancer Causes Control* **1997**, *8*, 420–443. [CrossRef] [PubMed]

25. Hossain, M.; Ngo, H.; Guo, W.; Nguyen, T.V.; Vigneswaran, S. Performance of cabbage and cauliflower wastes for heavy metals removal. *Desalin. Water Treat.* **2014**, *52*, 844–860. [CrossRef]

26. Dudgeon, D.; Arthington, A.H.; Gessner, M.O.; Kawabata, Z.-I.; Knowler, D.J.; Lévêque, C.; Naiman, R.J.; Prieur-Richard, A.-H.; Soto, D.; Stiassny, M.L.J.; et al. Freshwater biodiversity: Importance, threats, status and conservation challenges. *Biol. Rev.* **2006**, *81*, 163–182. [CrossRef] [PubMed]

27. Lintern, A.; Leahy, P.J.; Heijnis, H.; Zawadzki, A.; Gadd, P.; Jacobsen, G.; Deletic, A.; Mccarthy, D.T. Identifying heavy metal levels in historical flood water deposits using sediment cores. *Water Res.* **2016**, *105*, 34–46. [CrossRef]

28. Liu, M.; Du, P.; Yu, C.; He, Y.; Zhang, H.; Sun, X.; Lin, H.; Luo, Y.; Xie, H.; Guo, J.; et al. Increases of Total Mercury and Methylmercury Releases from Municipal Sewage into Environment in China and Implications. *Environ. Sci. Technol.* **2018**, *52*, 124–134. [CrossRef]

29. Crini, G. Non-conventional low-cost adsorbents for dye removal: A review. *Bioresour. Technol.* **2006**, *97*, 1061–1085. [CrossRef]

30. Daneshvar, N.; Oladegaragoze, A.; Djafarzadeh, N. Decolorization of basic dye solutions by electrocoagulation: An investigation of the effect of operational parameters. *J. Hazard. Mater.* **2006**, *129*, 116–122. [CrossRef]

31. Patil, Y.; Paknikar, K. Development of a process for biodetoxification of metal cyanides from waste waters. *Process. Biochem.* **2000**, *35*, 1139–1151. [CrossRef]

32. Nesbitt, A.B. Recovery of Metal Cyanides Using a Fluidized Bed of Resin. Master's Thesis, Cape Peninsula University of Technology, Cape Town, South Africa, December 1996.

33. Santos, B.A.Q.; Ntwampe, S.K.O.; Doughari, J.H.; Muchatibaya, G. Application of *Citrus sinensis* Solid Waste as a Pseudo-Catalyst for Free Cyanide Conversion under Alkaline Conditions. *Bioresources* **2013**, *8*, 3461–3467. [CrossRef]

34. Kostich, M.S.; Batt, A.L.; Lazorchak, J. Concentrations of prioritized pharmaceuticals in effluents from 50 large wastewater treatment plants in the US and implications for risk estimation. *Environ. Pollut.* **2014**, *184*, 354–359. [CrossRef]

35. Lessa, E.F.; Nunes, M.L.; Fajardo, A.R. Chitosan/waste coffee-grounds composite: An efficient and eco-friendly adsorbent for removal of pharmaceutical contaminants from water. *Carbohydr. Polym.* **2018**, *189*, 257–266. [CrossRef]

36. Larsson, D.G.J. Pollution from drug manufacturing: Review and perspectives. *Philos. Trans. R. Soc. B Biol. Sci.* **2014**, *369*, 20130571. [CrossRef]

37. Yan, Z.; Lu, G.; Liu, J.; Jin, S. An integrated assessment of estrogenic contamination and feminization risk in fish in Taihu Lake, China. *Ecotoxicol. Environ. Saf.* **2012**, *84*, 334–340. [CrossRef]

38. Adewuyi, A. Chemically modified bio-sorbents and their role in the removal of emerging pharmaceutical waste in the water system. *Water* **2020**, *12*, 1551. [CrossRef]

39. Swan, G.; Naidoo, V.; Cuthbert, R.; Green, R.E.; Pain, D.J.; Swarup, D.; Prakash, V.; Taggart, M.; Bekker, L.; Das, D.; et al. Removing the Threat of Diclofenac to Critically Endangered Asian Vultures. *PLoS Biol.* **2006**, *4*, e66. [CrossRef] [PubMed]

40. Oaks, J.L.; Gilbert, M.; Virani, M.Z.; Watson, R.T.; Meteyer, C.U.; Rideout, B.A. Diclofenac residues as the cause of vulture population decline in Pakistan. *Nature* **2004**, *427*, 630–633. [CrossRef]

41. Shultz, S.; Baral, H.S.; Charman, S.; Cunningham, A.A.; Das, D.; Ghalsasi, G.R.; Goudar, M.S.; Green, R.E.; Jones, A.; Nighot, P.; et al. Diclofenac poisoning is widespread in declining vulture populations across the Indian subcontinent. *Proc. R. Soc. B Boil. Sci.* **2004**, *271*, S458–S460. [CrossRef]

42. Green, R.E.; Newton, I.; Shultz, S.; Cunningham, A.A.; Pain, D.J.; Prakash, V. Diclofenac Poisoning as a Cause of Vulture Population Declines across the Indian Subcontinent. *J. Appl. Ecol.* **2004**, *41*, 793–800. [CrossRef]

43. Prakash, V.; Pain, D.; Cunningham, A.; Donald, P.; Verma, A.; Gargi, R.; Sivakumar, S.; Rahmani, A. Catastrophic collapse of Indian white-backed Gyps bengalensis and long-billed Gyps indicus vulture populations. *Biol. Conserv.* **2003**, *109*, 381–390. [CrossRef]

44. Patterson, J. *Industrial Wastewater Treatment Technology*, 2nd ed.; Patterson, J., Ed.; Butterworth Publishers: Stoneham, MA, USA, 1985.

45. Hameed, B.; Tan, I.A.W.; Ahmad, A.L. Optimization of basic dye removal by oil palm fibre-based activated carbon using response surface methodology. *J. Hazard. Mater.* **2008**, *158*, 324–332. [CrossRef]

46. Lata, H.; Garg, V.; Gupta, R. Removal of a basic dye from aqueous solution by adsorption using Parthenium hysterophorus: An agricultural waste. *Dye Pigment.* **2007**, *74*, 653–658. [CrossRef]

47. Tan, X.; Liu, Y.; Zeng, G.; Wang, X.; Hu, X.; Gu, Y.; Yang, Z. Application of biochar for the removal of pollutants from aqueous solutions. *Chemosphere* **2015**, *125*, 70–85. [CrossRef]

48. Shakoor, S.; Nasar, A. Removal of methylene blue dye from artificially contaminated water using *Citrus limetta* peel waste as a very low cost adsorbent. *J. Taiwan Inst. Chem. Eng.* **2016**, *66*, 154–163. [CrossRef]

49. World Health Organization. *Guidelines for Drinking—Water Quality*; WHO: Geneva, Switzerland, 2011.

50. Purkayastha, D.; Mishra, U.; Biswas, S. A comprehensive review on Cd(II) removal from aqueous solution. *J. Water Process. Eng.* **2014**, *2*, 105–128. [CrossRef]

51. Yadav, S.; Sinha, S.; Singh, D.K. Chromium(VI) removal from aqueous solution and industrial wastewater by modified date palm trunk. *Environ. Prog. Sustain. Energy* **2014**, *34*, 452–460. [CrossRef]

52. Bilal, M.; Shah, J.A.; Ashfaq, T.; Mubashar, S.; Gardazi, H.; Tahir, A.A. Waste biomass adsorbents for copper removal from industrial wastewater-A review. *J. Hazard. Mater.* **2013**, *263*, 322–333. [CrossRef]

53. Zahra, N. Lead Removal from Water by Low Cost Adsorbents: A Review. *Pak. J. Anal. Environ. Chem.* **2012**, *13*, 1–8.

54. Kanawade, S.M.; Gaikwad, R.W. Removal of Zinc Ions from Industrial Effluent by Using Cork Powder as Adsorbent. *Int. J. Chem. Eng. Appl.* **2011**, *2*, 199–201. [CrossRef]

55. Singh, T.P.; Majumder, C.B. Flouride removal from sewage water using *Citrus limetta* peel as bio-sorbent. *Int. J. Pharm. Pharm. Sci.* **2016**, *8*, 86–92.

56. Siddique, A.; Nayak, A.K.; Singh, J. Synthesis of FeCl3-activated carbon derived from waste *Citrus limetta* peels for removal of fluoride: An eco-friendly approach for the treatment of groundwater and bio-waste collectively. *Groundw. Sustain. Dev.* **2020**, *10*, 100339. [CrossRef]

57. Dolas, H.; Sahin, O.; Saka, C.; Demir, H. A new method on producing high surface area activated carbon: The effect of salt on the surface area and the pore size distribution of activated carbon prepared from pistachio shell. *Chem. Eng. J.* **2011**, *166*, 191–197. [CrossRef]

58. Özdemir, M.; Bolgaz, T.; Saka, C.; Şahin, Ö. Preparation and characterization of activated carbon from cotton stalks in a two-stage process. *J. Anal. Appl. Pyrolysis* **2011**, *92*, 171–175. [CrossRef]

59. Saka, C.; Sahin, O.; Celik, M.S. The Removal of Methylene Blue from Aqueous Solutions by Using Microwave Heating and Pre-boiling Treated Onion Skins as a New Adsorbent. *Energy Sources Part A Recover. Util. Environ. Eff.* **2012**, *34*, 1577–1590. [CrossRef]

60. Bulut, Y.; Aydın, H. A kinetics and thermodynamics study of methylene blue adsorption on wheat shells. *Desalination* **2006**, *194*, 259–267. [CrossRef]

61. Vadivelan, V.; Kumar, K.V. Equilibrium, kinetics, mechanism, and process design for the sorption of methylene blue onto rice husk. *J. Colloid Interface Sci.* **2005**, *286*, 90–100. [CrossRef]

62. Uddin, M.T.; Islam, M.A.; Mahmud, S.; Rukanuzzaman, M. Adsorptive removal of methylene blue by tea waste. *J. Hazard. Mater.* **2009**, *164*, 53–60. [CrossRef]

63. Annadurai, G.; Juang, R.-S.; Lee, D.-J. Use of cellulose-based wastes for adsorption of dyes from aqueous solutions. *J. Hazard. Mater.* **2002**, *92*, 263–274. [CrossRef]

64. Bhattacharya, K.G.; Sharma, A. Kinetics and thermodynamics of Methylene Blue adsorption on Neem (*Azadirachta indica*) leaf powder. *Dye Pigment.* **2005**, *65*, 51–59. [CrossRef]

65. McKay, G.; Porter, J.F.; Prasad, G.R. The Removal of Dye Colours from Aqueous Solutions by Adsorption on Low-cost Materials. *Water Air Soil Pollut.* **1999**, *114*, 423–438. [CrossRef]

66. Asgher, M.; Bhatti, H.N. Removal of reactive blue 19 and reactive blue 49 textile dyes by citrus waste biomass from aqueous solution: Equilibrium and kinetic study. *Can. J. Chem. Eng.* **2011**, *90*, 412–419. [CrossRef]

67. Bhatti, H.N.; Bajwa, I.I.; Hanif, M.A.; Bukhari, I.H. Removal of lead and cobalt using lignocellulosic fiber derived from *Citrus reticulata* waste biomass. *Korean J. Chem. Eng.* **2010**, *27*, 218–227. [CrossRef]

68. Won, S.W.; Kotte, P.; Wei, W.; Lim, A.; Yun, Y.S. Bio-sorbents for recovery of precious metals. *Bioresour. Technol.* **2014**, *160*, 203–212. [CrossRef]

69. Iqbal, M.; Schiewer, S.; Cameron, R. Mechanistic elucidation and evaluation of biosorption of metal ions by grapefruit peel using FTIR spectroscopy, kinetics and isotherms modeling, cations displacement and EDX analysis. *J. Chem. Technol. Biotechnol.* **2009**, *84*, 1516–1526. [CrossRef]

70. Suresh Babu, C.; Chakrapani, C.; Somasekhara Rao, K. Equilibrium and kinetic studies of reactive red 2 dye adsorption onto prepared activated carbons. *J. Chem. Pharm. Res.* **2011**, *3*, 428–439.

71. Chen, S.; Luo, J.; Hu, M.; Lai, K.; Geng, P.; Huang, H. Enhancement of cypermethrin degradation by a coculture of Bacillus cereus ZH-3 and Streptomyces aureus HP-S-01. *Bioresour. Technol.* **2012**, *110*, 97–104. [CrossRef]

72. Dutta, S.; Bhattacharyya, A.; Ganguly, A.; Gupta, S.; Basu, S. Application of Response Surface Methodology for preparation of low-cost adsorbent from citrus fruit peel and for removal of Methylene Blue. *Desalination* **2011**, *275*, 26–36. [CrossRef]

73. Mahmoud, G.A.; Abdel-Aal, S.E.; Badway, N.A.; Elbayaa, A.A.; Ahmed, D.F. A novel hydrogel based on agricultural waste for removal of hazardous dyes from aqueous solution and reuse process in a secondary adsorption. *Polym. Bull.* **2016**, *74*, 337–358. [CrossRef]

74. Rehman, R.; Zafar, J.; Nisar, H. Adsorption Studies of Removal of Indigo Caramine Dye from Water by Formaldehyde and Urea Treated Cellulosic Waste of *Citrus reticulata* Peels. *Asian J. Chem.* **2014**, *26*, 43–47. [CrossRef]

75. Rehman, R.; Salman, M.; Mahmud, T.; Kanwal, F.; Waheed-Uz-Zama. Utilization of chemically modified *Citrus reticulata* peels for biosorptive removal of acid yellow-73 dye from water. *J. Chem. Soc. Pak.* **2013**, *35*, 611–616.

76. Dev, S.; Khamkhash, A.; Ghosh, T.; Aggarwal, S. Adsorptive Removal of Se(IV) by Citrus Peels: Effect of Adsorbent Entrapment in Calcium Alginate Beads. *ACS Omega* **2020**, *5*, 17215–17222. [CrossRef] [PubMed]

77. Kam, S.K.; Lee, M.G. Adsorption characteristics of antibiotics amoxicillin in aqueous solution with activated carbon prepared from waste citrus peel. *Appl. Chem. Eng.* **2018**, *29*, 369–375.

78. Putra, E.K.; Pranowo, R.; Sunarso, J.; Indraswati, N.; Ismadji, S. Performance of activated carbon and bentonite for adsorption of amoxicillin from wastewater: Mechanisms, isotherms and kinetics. *Water Res.* **2009**, *43*, 2419–2430. [CrossRef]

79. Moussavi, G.; Alahabadi, A.; Yaghmaeian, K.; Eskandari, M. Preparation, characterization and adsorption potential of the NH4Cl-induced activated carbon for the removal of amoxicillin antibiotic from water. *Chem. Eng. J.* **2013**, *217*, 119–128. [CrossRef]

80. Ding, R.; Zhang, P.; Seredych, M.; Bandosz, T.J. Removal of antibiotics from water using sewage sludge- and waste oil sludge-derived adsorbents. *Water Res.* **2012**, *46*, 4081–4090. [CrossRef]

81. Baccar, R.; Sarra, M.; Bouzid, J.; Feki, M.; Blánquez, P. Removal of pharmaceutical compounds by activated carbon prepared from agricultural by-product. *Chem. Eng. J.* **2012**, *211–212*, 310–317. [CrossRef]

82. Ahmed, M.J.; Theydan, S.K. Microporous activated carbon from Siris seed pods by microwave-induced KOH activation for metronidazole adsorption. *J. Anal. Appl. Pyrolysis* **2013**, *99*, 101–109. [CrossRef]

83. Pouretedal, H.R.; Sadegh, N. Effective removal of Amoxicillin, Cephalexin, Tetracycline and Penicillin G from aqueous solutions using activated carbon nanoparticles prepared from vine wood. *J. Water Process. Eng.* **2014**, *1*, 64–73. [CrossRef]

84. Camel, V. Solid phase extraction of trace elements. *Spectrochim. Acta Part B At. Spectrosc.* **2003**, *58*, 1177–1233. [CrossRef]

85. Li, X.; Tang, Y.; Cao, X.; Lu, D.; Luo, F.; Shao, W. Preparation and evaluation of orange peel cellulose adsorbents for effective removal of cadmium, zinc, cobalt and nickel. *Colloids Surf. A Physicochem. Eng. Asp.* **2008**, *317*, 512–521. [CrossRef]

86. Sha, L.; Xueyi, G.; Ningchuan, F.; Qinghua, T. Adsorption of Cu^{2+} and Cd^{2+} from aqueous solution by mercapto-acetic acid modified orange peel. *Colloids Surf. B Biointerfaces* **2009**, *73*, 10–14. [CrossRef] [PubMed]

87. Kohn, R. Binding of divalent cations to oligomeric fragments of pectin. *Carbohydr. Res.* **1987**, *160*, 343–353. [CrossRef]

88. Malovíková, A.; Kohn, R. 702 Binding of Cadmium Cations to Pectin. *Collect. Czech. Chem. Commun.* **1982**, *47*, 702–708. [CrossRef]

89. Cameron, R.G.; Iqbal, M.; Widmer, W.; Haven, A.S.N.W.W. Biosorption Properties of Citrus Peel Derived Oligogalac-turonides, Enzyme-modified Pectin and Peel Hydrolysis Residues. *Proc. Fla. State Hortic. Soc.* **2008**, *5*, 311–314.

90. Haven, W. A Refereed Paper Production of Narrow-Range Size-Classes of Polygalacturonic Acid Oligomers. *Proc. Fla. State Hortic. Soc.* **2005**, *118*, 406–409.

91. Joye, D.; Luzio, G. Process for selective extraction of pectins from plant material by differential pH. *Carbohydr. Polym.* **2000**, *43*, 337–342. [CrossRef]

92. Arief, V.O.; Trilestari, K.; Sunarso, J.; Indraswati, N.; Ismadji, S. Recent progress on biosorption of heavy metals from liquids using low cost bio-sorbents: Characterization, biosorption parameters and mechanism studies. *Clean Soil Air Water* **2008**, *36*, 937–962. [CrossRef]

93. Vijayaraghavan, K.; Yun, Y.S. Bacterial bio-sorbents and biosorption. *Biotechnol. Adv.* **2008**, *26*, 266–291. [CrossRef] [PubMed]

94. Wang, J.; Chen, C. Biosorption of heavy metals by Saccharomyces cerevisiae: A review. *Biotechnol. Adv.* **2006**, *24*, 427–451. [CrossRef] [PubMed]

95. Mahvi, A.H. Application of agricultural fibers in pollution removal from aqueous solution. *Int. J. Environ. Sci. Technol.* **2008**, *5*, 275–285. [CrossRef]

96. Suryavanshi, U.; Shukla, S.R. Adsorption of Pb^{2+} by Alkali-Treated *Citrus limetta* Peels. *Ind. Eng. Chem. Res.* **2010**, *49*, 11682–11688. [CrossRef]

97. Chatterjee, A.; Schiewer, S. Biosorption of Cadmium(II) Ions by Citrus Peels in a Packed Bed Column: Effect of Process Pa-rameters and Comparison of Different Breakthrough Curve Models. *Clean Soil Air Water* **2011**, *39*, 874–881. [CrossRef]

98. Chatterjee, A.; Schiewer, S. Effect of Competing Cations (Pb, Cd, Zn, and Ca) in Fixed-Bed Column Biosorption and Desorption from Citrus Peels. *Water Air Soil Pollut.* **2014**, *225*, 1854. [CrossRef]

99. Ghimire, K.N.; Inoue, K.; Yamaguchi, H.; Makino, K.; Miyajima, T. Adsorptive separation of arsenate and arsenite anions from aqueous medium by using orange waste. *Water Res.* **2003**, *37*, 4945–4953. [CrossRef]

100. Gondhalekar, S.C.; Shukla, S.R. Equilibrium and kinetics study of uranium(VI) from aqueous solution by *Citrus limetta* peels. *J. Radioanal. Nucl. Chem.* **2014**, *302*, 451–457. [CrossRef]

101. Khaskheli, M.I.; Memon, S.Q.; Parveen, S.; Khuhawar, M.Y. *Citrus paradisi*: An Effective bio-adsorbent for Arsenic(V) Remediation. *Pak. J. Anal. Environ. Chem.* **2014**, *15*, 35–41.

102. Khaskheli, M.I.; Memon, S.Q.; Siyal, A.N.; Khuhawar, M.Y. Use of Orange Peel Waste for Arsenic Remediation of Drinking Water. *Waste Biomass Valoriz.* **2011**, *2*, 423–433. [CrossRef]

103. Gönen, F. Adsorption study on orange peel: Removal of Ni(II) ions from aqueous solution. *Afr. J. Biotechnol.* **2012**, *11*, 1250–1258. [CrossRef]

104. Marín, A.B.P.; Aguilar, M.I.; Ortuño, J.F.; Meseguer, V.F.; Sáez, J.; Llorens, M. Biosorption of Zn(II) by orange waste in batch and packed-bed systems. *J. Chem. Technol. Biotechnol.* **2010**, *85*, 1310–1318. [CrossRef]

105. Pavan, F.A.; Lima, I.S.; Lima, É.C.; Airoldi, C.; Gushikem, Y. Use of Ponkan mandarin peels as bio-sorbent for toxic metals uptake from aqueous solutions. *J. Hazard. Mater.* **2006**, *137*, 527–533. [CrossRef]

106. Abdul Sattar, J.A. Toxic Metal Pollution Abatement Using Sour Orange Biomass. *J. Al-Nahrain Univ. Sci.* **2013**, *16*, 56–64. [CrossRef]

107. Torab-Mostaedi, M. Biosorpcija lantana i cerijuma iz vodenih rastvora pomoću kore mandarine (*Citrus reticulata*): Ravnotežna, kinetička i termodinamička ispitivanja. *Chem. Ind. Chem. Eng. Q.* **2013**, *19*, 79–88. [CrossRef]

108. Pavan, F.A.; Mazzocato, A.C.; Jacques, R.A.; Dias, S.L.P. Ponkan peel: A potential bio-sorbent for removal of Pb(II) ions from aqueous solution. *Biochem. Eng. J.* **2008**, *40*, 357–362. [CrossRef]

109. Pérez-Marín, A.; Ballester, A.; González, F.; Blázquez, M.; Muñoz, J.; Sáez, J.; Zapata, V.M. Study of cadmium, zinc and lead biosorption by orange wastes using the subsequent addition method. *Bioresour. Technol.* **2008**, *99*, 8101–8106. [CrossRef]

110. Rane, N.M.; Sapkal, R.S. Chromium(Vi) Removal By Using Orange Peel Powder in Batch Adsorption. *Int. J. Chem. Sci. Appl.* **2014**, *5*, 2278–6015.

111. Santos, C.M.; Dweck, J.; Viotto, R.S.; Rosa, A.H.; de Morais, L.C. Application of orange peel waste in the production of solid biofuels and bio-sorbents. *Bioresour. Technol.* **2015**, *196*, 469–479. [CrossRef]

112. Njikam, E.; Schiewer, S. Optimization and kinetic modeling of cadmium desorption from citrus peels: A process for bio-sorbent regeneration. *J. Hazard. Mater.* **2012**, *213–214*, 242–248. [CrossRef] [PubMed]

113. Sowmya Lakshmi, K.B.; Munilakshmi, N. Adsorptive removal of colour from aqueous solution of disazo dye by using organic adsorbents. *Int. J. Chem. Technol. Res.* **2016**, *9*, 407–415.

114. Lasheen, M.R.; Ammar, N.S.; Ibrahim, H.S. Adsorption/desorption of Cd(II), Cu(II) and Pb(II) using chemically modified orange peel: Equilibrium and kinetic studies. *Solid State Sci.* **2012**, *14*, 202–210. [CrossRef]

115. Balaria, A.; Schiewer, S. Assessment of biosorption mechanism for Pb binding by citrus pectin. *Sep. Purif. Technol.* **2008**, *63*, 577–581. [CrossRef]

116. Biswas, B.K.; Inoue, K.; Ghimire, K.N.; Ohta, S.; Harada, H.; Ohto, K.; Kawakita, H. The adsorption of phosphate from an aquatic environment using metal-loaded orange waste. *J. Colloid Interface Sci.* **2007**, *312*, 214–223. [CrossRef]

117. Feng, N.-C.; Guo, X. Characterization of adsorptive capacity and mechanisms on adsorption of copper, lead and zinc by modified orange peel. *Trans. Nonferrous Met. Soc. China* **2012**, *22*, 1224–1231. [CrossRef]

118. Mandina, S. Removal of chromium (VI) from aqueous solution using chemically modified orange (citrus cinensis) peel. *IOSR J. Appl. Chem.* **2013**, *6*, 66–75. [CrossRef]

119. Bhatnagar, A.; Minocha, A.K.; Sillanpää, M. Adsorptive removal of cobalt from aqueous solution by utilizing lemon peel as bio-sorbent. *Biochem. Eng. J.* **2010**, *48*, 181–186. [CrossRef]

120. De Souza, J.V.T.M.; Diniz, K.M.; Massocatto, C.L.; Tarley, C.R.T.; Caetano, J.; Dragunski, D.C. Removal of Pb (II) from aqueous solution with orange sub-products chemically modified as biosorbent. *BioResources* **2012**, *7*, 2300–2318. [CrossRef]

121. Tasaso, P. Adsorption of Copper Using Pomelo Peel and Depectinated Pomelo Peel. *J. Clean Energy Technol.* **2014**, *2*, 154–157. [CrossRef]

122. Schiewer, S.; Iqbal, M. The role of pectin in Cd binding by orange peel bio-sorbents: A comparison of peels, depectinated peels and pectic acid. *J. Hazard. Mater.* **2010**, *177*, 899–907. [CrossRef] [PubMed]

123. Schiewer, S.; Patil, S.B. Modeling the effect of pH on biosorption of heavy metals by citrus peels. *J. Hazard. Mater.* **2008**, *157*, 8–17. [CrossRef]

124. Singh, S.; Shukla, S.R. Adsorptive removal of cobalt ions on raw and alkali-treated lemon peels. *Int. J. Environ. Sci. Technol.* **2015**, *13*, 165–178. [CrossRef]

125. Crockett, M.; Evans, A.; Worthington, M.J.H.; Albuquerque, I.S.; Slattery, A.D.; Gibson, C.; Campbell, J.; Lewis, D.A.; Bernardes, G.; Chalker, J.M. Sulfur-Limonene Polysulfide: A Material Synthesized Entirely from Industrial By-Products and Its Use in Removing Toxic Metals from Water and Soil. *Angew. Chem. Int. Ed.* **2016**, *55*, 1714–1718. [CrossRef] [PubMed]

126. Namasivayam, C.; Muniasamy, N.; Gayatri, K.; Rani, M.; Ranganathan, K. Removal of dyes from aqueous solutions by cellulosic waste orange peel. *Bioresour. Technol.* **1996**, *57*, 37–43. [CrossRef]
127. Khaled, A.; El Nemr, A.; El-Sikaily, A.; Abdelwahab, O. Treatment of artificial textile dye effluent containing Direct Yellow 12 by orange peel carbon. *Desalination* **2009**, *238*, 210–232. [CrossRef]
128. Saeed, A.; Sharif, M.; Iqbal, M. Application potential of grapefruit peel as dye sorbent: Kinetics, equilibrium and mechanism of crystal violet adsorption. *J. Hazard. Mater.* **2010**, *179*, 564–572. [CrossRef]
129. Foo, K.Y.; Hameed, B.H. Microwave assisted preparation of activated carbon from pomelo skin for the removal of anionic and cationic dyes. *Chem. Eng. J.* **2011**, *173*, 385–390. [CrossRef]
130. Jayarajan, M.; Arunachala, R.; Annadurai, G. Use of Low Cost Nano-porous Materials of Pomelo Fruit Peel Wastes in Removal of Textile Dye. *Res. J. Environ. Sci.* **2011**, *5*, 434–443. [CrossRef]
131. Bhatnagar, A.; Kumar, E.; Minocha, A.K.; Jeon, B.-H.; Song, H.; Seo, Y.-C. Removal of Anionic Dyes from Water usingCitrus limonum(Lemon) Peel: Equilibrium Studies and Kinetic Modeling. *Sep. Sci. Technol.* **2009**, *44*, 316–334. [CrossRef]
132. Bhatti, H.N.; Akhtar, N.; Saleem, N. Adsorptive Removal of Methylene Blue by Low-Cost *Citrus sinensis* Bagasse: Equilibrium, Kinetic and Thermodynamic Characterization. *Arab. J. Sci. Eng.* **2011**, *37*, 9–18. [CrossRef]
133. Kučić, D.; Miljanić, S.; Rožić, M. Sorption of Methylene Blue onto Orange and Lemon Peel. *Holist. Approach Environ.* **2011**, *1*, 39–49.
134. Rosales, E.; Meijide, J.; Tavares, T.; Pazos, M.; Sanromán, M.A. Grapefruit peelings as a promising bio-sorbent for the removal of leather dyes and hexavalent chromium. *Process. Saf. Environ. Prot.* **2016**, *101*, 61–71. [CrossRef]
135. Hameed, B.H.; Mahmoud, D.; Ahmad, A.L. Sorption of basic dye from aqueous solution by pomelo (*Citrus grandis*) peel in a batch system. *Colloids Surf. A Physicochem. Eng. Asp.* **2008**, *316*, 78–84. [CrossRef]
136. Patil, P.R. Adsorption of Erichrome black T from aqueous solutions on activated carbon prepared from mosambi peel. *J. Appl. Sci. Environ. Sanit.* **2014**, *6*, 149–154.
137. Mafra, M.R.; Igarashimafra, L.; Zuim, D.R.; Vasques, É.C.; Ferreira, M.A. Adsorption of remazol brilliant blue on an orange peel adsorbent. *Braz. J. Chem. Eng.* **2013**, *30*, 657–665. [CrossRef]
138. Pavlović, M.D.; Nikolić, I.R.; Milutinović, M.D.; Dimitrijević-Branković, S.I.; Šiler-Marinković, S.S.; Antonović, D.G. Iskorišćenje sirovog otpada iz restorana za adsorpciju boja. *Hem. Ind.* **2015**, *69*, 667–677.
139. Rahmat, N.A.; Ali, A.A.; Salmiati; Hussain, N.; Muhamad, M.S.; Kristanti, R.; Hadibarata, T. Removal of Remazol Brilliant Blue R from Aqueous Solution by Adsorption Using Pineapple Leaf Powder and Lime Peel Powder. *Water Air Soil Pollut.* **2016**, *227*, 1–11. [CrossRef]
140. Bello, O.S.; Ahmad, M.A.; Semire, B. Scavenging malachite green dye from aqueous solutions using pomelo (*Citrus grandis*) peels: Kinetic, equilibrium and thermodynamic studies. *Desalin. Water Treat.* **2015**, *56*, 521–535. [CrossRef]
141. Sivaraj, R.; Namasivayam, C.; Kadirvelu, K. Orange peel as an adsorbent in the removal of Acid violet 17 (acid dye) from aqueous solutions. *Waste Manag.* **2001**, *21*, 105–110. [CrossRef]
142. Salman, T.A.; Ali, M.I. Potential Application of Natural and Modified Orange Peel as an Eco-friendly Adsorbent for Methylene Blue Dye. *Iraqi J. Sci.* **2016**, *57*, 1–13.
143. Lagergren, S. Zur Theorie der Sogenannten Adsorption Gelöster Stoffe. *Sven. Vetensk. Handl.* **1998**, *24*, 1–39.
144. Ho, Y.S.; McKay, G. Pseudo-second order model for sorption processes. *Process. Biochem.* **1999**, *34*, 451–465. [CrossRef]
145. Shakoor, S.; Nasar, A. Adsorptive treatment of hazardous methylene blue dye from artificially contaminated water using cucumis sativus peel waste as a low-cost adsorbent. *Groundw. Sustain. Dev.* **2017**, *5*, 152–159. [CrossRef]
146. Levankumar, L.; Muthukumaran, V.; Gobinath, M. Batch adsorption and kinetics of chromium (VI) removal from aqueous solutions by *Ocimum americanum* L. seed pods. *J. Hazard. Mater.* **2009**, *161*, 709–713. [CrossRef]
147. Elovich, S.Y.; Larinov, O.G. Theory of Adsorption from Solutions of Non Electrolytes on Solid (I) Equation Adsorption from Solutions and the Analysis of Its Simplest Form, (II) Verification of the Equation of Adsorption Isotherm from Solutions. *Izv. Akad. Nauk. SSSR* **1962**, *2*, 209–216.
148. Weber, W.J.; Morris, J.C. Advances in water pollution research: Removal of biologically resistant pollutant from waste water by adsorption. In *Proceedings of the 1st International Conference on Water Pollution Symposium*; Pregamon Press: Oxford, UK, 1962; Volume 2, pp. 231–266.
149. Tan, I.A.W.; Hameed, B.H.; Ahmad, A.L. Equilibrium and kinetic studies on basic dye adsorption by oil palm fibre activated carbon. *Chem. Eng. J.* **2007**, *127*, 111–119. [CrossRef]
150. Freundlich, H.M.F. Over the adsorption in solution. *J. Phys. Chem.* **1906**, *57*, 385–471.
151. Wijaya, R.; Andersan, G.; Santoso, S.P.; Irawaty, W. Green Reduction of Graphene Oxide using Kaffir Lime Peel Extract (*Citrus hystrix*) and Its Application as Adsorbent for Methylene Blue. *Sci. Rep.* **2020**, *10*, 1–9. [CrossRef]
152. Mondal, N.K.; Basu, S.; Sen, K.; Debnath, P. Potentiality of mosambi (*Citrus limetta*) peel dust toward removal of Cr(VI) from aqueous solution: An optimization study. *Appl. Water Sci.* **2019**, *9*, 116. [CrossRef]
153. Ahmed, M.; Mashkoor, F.; Nasar, A. Development, characterization, and utilization of magnetized orange peel waste as a novel adsorbent for the confiscation of crystal violet dye from aqueous solution. *Groundw. Sustain. Dev.* **2020**, *10*, 100322. [CrossRef]
154. Sharifzade, G.; Asghari, A.; Rajabi, M. Highly effective adsorption of xanthene dyes (rhodamine B and erythrosine B) from aqueous solutions onto lemon citrus peel active carbon: Characterization, resolving analysis, optimization and mechanistic studies. *RSC Adv.* **2017**, *7*, 5362–5371. [CrossRef]

155. Jawad, A.H.; Al-Heetimi, D.T.A.; Mastuli, M.S. Biochar from orange (*Citrus sinensis*) peels by acid activation for methylene blue adsorption. *Iran J. Chem. Chem. Eng.* **2019**, *38*, 91–105.
156. López-Manchado, M.; Arroyo, M. Effect of the incorporation of pet fibers on the properties of thermoplastic elastomer based on PP/elastomer blends. *Polymer* **2001**, *42*, 6557–6563. [CrossRef]
157. Ziabari, M.; Mottaghitalab, V.; Haghi, A.K. A new approach for optimization of electrospun nanofiber formation process. *Korean J. Chem. Eng.* **2010**, *27*, 340–354. [CrossRef]
158. Agarwal, P.; Mishra, P.K.; Srivastava, P. Statistical optimization of the electrospinning process for chitosan/polylactide nanofabrication using response surface methodology. *J. Mater. Sci.* **2012**, *47*, 4262–4269. [CrossRef]
159. Agarwal, P.; Mondal, A.; Mishra, P.; Srivastava, P. A predictive model for the synthesis of polylactide from lactic acid by response surface methodology. *ePolymers* **2011**, *11*, 739. [CrossRef]
160. Sudamalla, P.; Pichiah, S.; Manickam, M. Responses of surface modeling and optimization of Brilliant Green adsorption by adsorbent prepared from *Citrus limetta* peel. *Desalin. Water Treat.* **2012**, *50*, 367–375. [CrossRef]
161. Kam, S.K.; Lee, M.G. Response surface modeling for the adsorption of dye eosin Y by activated carbon prepared from waste citrus peel. *Appl. Chem. Eng.* **2018**, *29*, 270–277.
162. Lee, M.G.; Kam, S.K. Study on the adsorption of antibiotics trimethoprim in aqueous solution by activated carbon prepared from waste citrus peel using box-behnken design. *Korean Chem. Eng Res.* **2018**, *56*, 568–576.

MDPI
St. Alban-Anlage 66
4052 Basel
Switzerland
Tel. +41 61 683 77 34
Fax +41 61 302 89 18
www.mdpi.com

Processes Editorial Office
E-mail: processes@mdpi.com
www.mdpi.com/journal/processes